The Future of Food

PIONEERS

AI can bring significant transformation in the livestock sector by supporting animal health surveillance, early detection of transboundary and endemic diseases, precision nutrition, genetic improvement, reproductive management, and welfare monitoring.

Dr Swati Dahiya,
Lala Lajpat Rai University of Veterinary
and Animal Sciences, Hisar

AI chatbots provide multilingual, voice-based advice on wheat/barley queries, drawing from Kisan Call Centre data to address climate, stress, and market issues for Indian farmers. This fosters trust and scales knowledge dissemination ethically.

Dr Ratan Tiwari,
Director, ICAR-Indian Institute of
Wheat and Barley Research, Karnal

As agricultural leadership goes beyond mere technical proficiency in a particular domain, embracing humanities, science, and AI in agricultural leadership is a revolutionary step towards transforming agriculture & allied sectors.

Dr N. Felix,
Former Vice Chancellor, Tamil Nadu
Dr J. Jayalalithaa Fisheries University

To maximise the benefits of technologies, particularly AI, the leadership in agriculture needs a bottom-up approach for fully grasping the socioeconomic realities to ensure that the potential of this technology does not bypass them. It, therefore, requires a seamless blending of the humanities (e.g., community engagement), scientific acumen (e.g., data-driven decision-making), and technological awareness (e.g., AI, blockchain, digital literacy). This will guarantee equity and avoid exclusivity.

Dr A. K. Singh,
Former Deputy Director General
(Natural Resource Management), ICAR

This vision is timely and commendable, and I welcome the effort to create a holistic leadership model that truly empowers smallholder farmers.

Dr I. J. Mathur,
Former, DEE, Maharana Pratap
University of Agriculture and Technology, Udaipur

Agriculture has always been more than the science of growing crops—it is the art of nurturing life itself. Today, as we face the twin challenges of climate change and food security, we must become bridge-builders: joining the precision of science with the empathy of the humanities, and letting AI serve as the bridge between them. Just as a farmer blends soil, seed, and season to bring forth a harvest, our leaders must blend technology, tradition, and human values to cultivate a future that is sustainable, inclusive, and just for all.

Dr G. R. Chintala,
Former Chairman, NABARD

AI serves as a bridge between data-driven science and human-centred values, while emphasising sustainability, inclusivity and the ethical dimension of leadership. This initiative has strong potential to be a valuable resource for policymakers, researchers and practitioners.

Dr Kuruba Gopal,
Vice-Chancellor, Dr YSR Horticultural University,
Tadepalligudem, West Godavari

AI is an emerging technology for precision agriculture, which is inevitable if a country has to become a developed nation. Machine learning, real-time data and decision support are done through AI integration.

Dr H. P. Singh,
Former DDG (Horticulture), ICAR and Former Vice-Chancellor,
Dr RPAU, Samastipur

Successful agricultural leadership requires a mindset that embraces change, encourages interdisciplinary collaboration, and views technology as a tool to serve human and environmental needs, not an end in itself. The future of food security, the environment, and rural livelihoods depends on leaders who can master this imperative blend.

Dr N. S. Rathore,
Former DDG and Former
Vice-Chancellor MPUAT, Udaipur

From a fisheries perspective, AI holds immense potential in areas such as stock prediction, habitat mapping, sustainable harvest optimisation, biodiversity monitoring, and value-chain enhancement.

Dr Grinson George,
Director, ICAR-CMFRI, Kochi

We have an Automation and Sensor Technology Division that focuses on automating post-harvest operations to enhance control, efficiency, and affordability, while also developing sensor-based quality assessment tools for food safety. The division's key thrust areas include integrating automation and sensors into existing machines and adapting technologies for local needs. We are eager to contribute to this collaborative journey of integrating AI as a powerful enabler for smallholder Indian farmers.

Dr Nachiket Kotwaliwale,
Director, Central Institute of Post-Harvest
Engineering & Technology, Ludhiana

This book holds significant value for a broad range of stakeholders: it aids farmers in boosting productivity without compromising culture, ethics, or environmental health; equips students to meld technical acumen with historical insight, policy awareness, and human values; assists researchers in crafting studies that are inclusive, just, and socially sensitive; and assists policymakers in formulating programmes and regulations that safeguard rights, ensure fair distribution of benefits, conserve ecosystems and cultural heritage, while fostering innovation and resilience.

Dr J. P. Mishra,
Director, ICAR-Agricultural Technology
Application Research Institute, Jodhpur

The Imperative Blend: Humanities, Science, and AI in Agricultural Leadership offers a timely and thought-provoking perspective on the future of agriculture. By framing artificial intelligence as both a tool and a bridge between empirical knowledge and human values, the book encourages researchers, emerging scientists, and policymakers to rethink agricultural priorities and promote inclusive, grassroots-oriented governance. Through its interdisciplinary lens, this book charts a path toward an agricultural future that is not only innovative and data-driven but also socially responsible and deeply attuned to community well-being.

Dr V. P. Chovatia,
Vice Chancellor, Junagadh
Agricultural University, Junagadh

The theme of this initiative is indeed most relevant and inspiring, particularly in the Indian context, where the integration of AI with agricultural research and leadership holds immense promise for transforming the livelihoods of smallholder farmers.

Dr Dheer Singh,
Director and Vice-Chancellor, ICAR-National
Dairy Research Institute, Karnal

The Future of Food

Agriculture Leadership in Tech-driven World

Dr Gopal Lal,
Arun Tiwari

Sakāl
Publications

Sakal Publications

THE FUTURE OF FOOD

Sakal Media Pvt. Ltd.
595, Budhwar Peth,
Pune–411002, India

www.sakalmediagroup.com
sakalpublications.com
sakalprakashan@esakal.com

First Edition: November, 2025

ISBN: 978-93-49487-27-7

Proofread by: Sneha Makharia
Typesetting: MAP Systems, Bengaluru

Printed in India by Sakal Media Pvt. Ltd.

A guide and a call to action for the next generation of agricultural scientists and agribusiness leaders. It demonstrates how Artificial Intelligence, when combined with ethics, innovation, and interdisciplinary knowledge, can empower smallholders, reshape food systems, and unlock opportunities that extend beyond conventional science.

Contents

Foreword

Agriculture has always been more than the act of producing food—it is the art of sustaining life, nurturing civilisations, and shaping the destiny of nations. As we step into the 21st century, the challenges facing agriculture are profound: climate change, shrinking natural resources, rapid urbanisation, and the mounting responsibility of feeding a growing population. Yet within these challenges lie unprecedented opportunities, driven by knowledge, technology, and above all, leadership.

This book, *The Future of Food: Agriculture Leadership in a Tech-driven World*, arrives at a most critical juncture. It is not merely a compilation of ideas but a call to action, designed to inspire and guide the next generation of agricultural scientists, policymakers, and entrepreneurs. Through its chapters, it highlights the centrality of smallholders—who form the backbone of global food systems—and demonstrates how innovation must serve their empowerment.

This volume is a timely contribution to the discourse on agricultural transformation. It urges us to think not only about "the future we want" but also about the responsibilities we must embrace to achieve it. For India, and indeed for the world, the message is clear: agriculture must remain values-driven, inclusive, and technologically forward-looking.

I commend the authors for this compilation. May this book inspire many young minds to take up the torch, to dream boldly, and to lead agriculture into a future where technology and humanity walk hand in hand.

New Delhi
October 2, 2025

Dr Mangi Lal Jat,
Secretary DARE, Govt of India,
Director General, ICAR

Introduction

Agriculture has always been more than an industry. It is humanity's oldest science, its most enduring art, and the foundation upon which civilisations have risen. Yet in the 21st century, agriculture faces unprecedented challenges—feeding a growing population, preserving fragile ecosystems, and creating livelihoods in a rapidly changing climate. At the same time, we stand on the brink of an extraordinary opportunity: a new age of Artificial Intelligence (AI) and digital transformation that can empower leaders to reshape the way food is grown, shared, and valued.

This book is written for the next generation of agricultural scientists, innovators, and agribusiness leaders—the young minds who will inherit the responsibility of feeding the world. Its central message is urgent yet straightforward: we cannot afford to evade AI. Instead, we must embrace it with vision, ethics, and courage. Technology alone will not solve our problems, but when put into the hands of purposeful leaders, it can become a force for empowerment, equity, and sustainability.

We begin by examining the transformations already reshaping our fields and markets. In the first two chapters, which set the context, from climate volatility to shifting consumer expectations, today's leaders must grasp the complexity of the modern food system. We focus on the millions of farmers who form the backbone of global agriculture. Their challenges—and their resilience—reveal why inclusive innovation matters most.

Chapter 3 invites readers to think past conventional boundaries. AI is not just another scientific tool; it is a bridge connecting biology, economics, social science, and ethics into a new kind of problem-solving. Chapter 4 explores how digital tools can be designed not for profit alone, but for empowerment—ensuring that progress reaches the farmer in the remotest village as much as it does the global agribusiness boardroom.

With Chapter 5, we turn to the critical task of reimagining farmer education and outreach. AI-enabled advisory systems, digital platforms, and new channels of communication can create a future where knowledge flows seamlessly to those who need it most. But technology without values can be dangerous. The chapter on Values-Driven Leadership underscores the importance of ethical responsibility in the digital transformation of agriculture. Leaders must ensure transparency, accountability, and fairness in every application of AI.

The following two chapters focus on harnessing Artificial Intelligence, exploring practical pathways that demonstrate how precision agriculture, risk management, automation, and data-driven decision-making can enhance both productivity and resilience, the skills young leaders must cultivate: interdisciplinary thinking, adaptability, and the capacity to inspire. Leadership in the AI age demands more than technical knowledge; it requires empathy, vision, and the ability to guide others through uncertainty.

The next two chapters lay out a roadmap for turning ideas into action—whether through startups, cooperatives, research initiatives, or policy advocacy. These pathways are designed to show

how small steps can lead to systemic change. We invoke reflection by asking what kind of world we want to build with the tools now in our hands? Will we create a future where food systems sustain people and planet, or one where technology deepens inequality? The answer depends on the choices made by today's young leaders.

In the last two chapters, we pay reverent tribute to the guides and visionaries who transformed agriculture from a fragmented practice into a rigorous, interdisciplinary science with a compassionate human face. These remarkable individuals ignited in generations of young people a lifelong commitment to feeding humanity, caring for livestock, nurturing soils, and safeguarding the natural world. It is because of their clarity of purpose, moral courage, and quiet sacrifices that Indian agriculture today stands as a beacon of self-reliance and resilience. Their legacy continues to illuminate the path for all who work to heal the Earth and nourish its people.

This book is not just a collection of ideas; it is an invitation to explore. An invitation to embrace AI as a partner in reimagining agriculture, to lead with values and courage, and to build food systems that are not only productive, but also just, inclusive, and sustainable. The future of food belongs to those who dare to lead it.

Hyderabad G.L.
October 2025 A.T.

1

Agriculture in the 21st Century

> ***The future of food will not be built by those who grow more, but by those who grow differently—leaders who cultivate not just crops, but sustainability, equity, and hope.***

1.1 Paradigm Shift

The evolution of agriculture from mere food production to environmental stewardship represents a paradigm shift in how humans interact with and manage natural resources. This transformation can be understood through three phases: pre-mechanisation (Pre-20th Century), the Industrial Revolution and Green Revolution (Mid-20th Century), and the Recognition of Environmental Impacts (Late 20th Century and into the 21st Century).

Initially, agriculture was primarily focused on maximising food and raw material output to support expanding populations and emerging civilisations. Practices during this period were often traditional, relying on local knowledge, and lacked a systematic understanding of ecological impacts. The advent of mechanisation, chemical fertilisers, pesticides, and high-yield crop varieties dramatically increased productivity. While these innovations helped combat food shortages, they also led to environmental issues, including soil degradation, water contamination, and biodiversity loss.

Growing awareness of the adverse effects of intensive farming practices prompted calls for more sustainable approaches. Issues such as pesticide runoff, monoculture vulnerability, and soil erosion have become prominent, highlighting the need for a balance between productivity and environmental health. Today, agriculture increasingly incorporates principles of ecological stewardship. It has already been identified as enhancing soil health and reducing erosion through crop rotation and cover cropping, integrated pest management (IPM), water conservation technologies, and organic farming. Emerging technologies, such as precision agriculture, remote sensing, and AI, help optimise resource use, reducing waste and environmental impact, and enable farmers to make data-informed decisions that support sustainable practices.

Overall, the shift from solely focusing on food production to embracing environmental stewardship reflects a holistic understanding that healthy ecosystems are crucial for long-term agricultural productivity. This integrated approach aims to achieve food security, economic viability, and ecological resilience, ensuring the well-being of both people and the planet for generations to come. This has led to a new type of leadership often referred to as sustainable leadership or stewardship leadership. We call it the Green Leadership. The shift from focusing only on food production to embracing environmental stewardship marks the rise of Green Leadership. This holistic vision recognises healthy ecosystems as the foundation of lasting agricultural productivity. By uniting food security, economic viability, and environmental resilience, Green Leadership ensures the well-being of people and the planet for generations to come.

1.2 Green Leadership

This form of leadership emphasises a holistic approach that balances economic, social, and environmental considerations. Key characteristics of this new leadership include visionary and long-term thinking, responsibility and accountability, collaborative and inclusive, innovation and adaptability, ethical and value-driven, and education and advocacy.

Green leadership is stewardship—rooted in ethics, it unites respect for nature, justice for people, and responsibility for future generations.

Green leaders prioritise the future health of ecosystems and communities, making decisions that benefit both present and future generations. They take responsibility for their impact on the planet, ensuring transparency and accountability in their actions. They create collaboration among diverse stakeholders—farmers, scientists, policymakers, communities, and companies—to develop sustainable solutions. They also foster innovation and embrace new technologies and practices that promote sustainability while adapting to changing environmental conditions.

Green Leadership is rooted in ethics, emphasising respect for nature, social justice, and the well-being of all living beings. A part of this leadership involves educating others about sustainability challenges and advocating for policies and practices that protect the environment. It is no wonder that farmers and industry leaders now view themselves as stewards of the land, responsible for maintaining healthy ecosystems. Working across sectors and disciplines becomes essential to develop sustainable agricultural systems. Green leaders influence policies that promote environmental conservation and responsible land use.

This new leadership paradigm represents nothing less than a renaissance of responsibility. It is born from the awareness that humanity's destiny is inextricably linked to the Earth's destiny. Leaders who embrace ecological consciousness are not merely managers of systems; they are guardians of life itself. Their decisions are guided by an ethic of care—for soil, water, air, and all living beings—because they understand that prosperity cannot be built on depletion.

They nurture a culture of innovation rooted in ethics, where technology serves as a tool for sustainability rather than an instrument of unchecked exploitation. They inspire collaboration not only through authority, but also through shared purpose, building bridges between farmers and technologists, policymakers and communities, and citizens and consumers. In this paradigm, leadership becomes a collective act of stewardship, carrying forward the profound truth that we borrow the Earth from our children, not inherit it from our ancestors.

Green Leadership is characterised by a holistic and future-oriented approach that balances economic, social, and environmental goals. Key traits include visionary long-term thinking, a deep sense of responsibility and accountability, and a commitment to transparency. Green leaders foster collaboration across diverse stakeholders—farmers, scientists, policymakers, and communities—working together to develop sustainable solutions.

They emphasise innovation and adaptability, embracing new technologies and practices that promote environmental health while responding to changing conditions. Rooted in strong ethical principles, Green Leadership values respect for nature, social justice, and the welfare of all living beings. Educating others about sustainability challenges and advocating for environmentally responsible policies are central to their work. These leaders view themselves as stewards of the land, dedicated to preserving healthy ecosystems and promoting sustainable land use. By integrating ecological consciousness into decision-making, Green Leaders cultivate a culture of responsibility, innovation, and collaboration, recognising that true sustainability requires committed environmental stewardship.

1.3 The Triple Challenges

Environmental degradation, climate change, and economic inequalities represent the three interconnected and pressing challenges that humanity must address to ensure sufficient food for all. Ecological degradation, through deforestation, soil erosion, and water pollution, diminishes the natural resources vital for productive agriculture. Climate change exacerbates these issues by causing unpredictable weather patterns, droughts, and floods, which threaten crop yields and food security worldwide. Simultaneously, economic inequalities restrict access to land, resources, and opportunities for marginalised populations, making it difficult for many farmers to adopt sustainable practices or improve productivity. Together, these triple challenges undermine global efforts to grow enough food, highlighting the urgent need for integrated solutions that promote environmental sustainability, climate resilience, and social equity.

Environmental degradation refers to the deterioration of natural ecosystems resulting from activities such as deforestation, soil erosion, water pollution, and loss of biodiversity. It reduces the availability of healthy soils, clean water, and biodiversity, which are essential for resilient agricultural systems. The examples include over-farming, excessive use of chemical fertilisers and pesticides, and deforestation for agriculture, which lead to soil fertility loss and habitat destruction, ultimately decreasing crop yields and livestock health.

Climate change involves long-term shifts in weather patterns caused by greenhouse gas emissions, primarily from human activities such as fossil fuel burning and deforestation.

It results in unpredictable weather, droughts, floods, and rising temperatures, disrupting planting and harvesting cycles. Examples include increased frequency of droughts reducing water availability for irrigation, floods destroying crops, and changing rainfall patterns, making farming more unpredictable, and threatening food security globally.

Economic inequalities refer to disparities in income, resources, and access to opportunities among different populations. They limit the ability of smallholder farmers and marginalised communities to adopt sustainable practices or access inputs like seeds, fertilisers, and technology. Examples include poor farmers lacking access to credit or markets, leading to over-reliance on subsistence farming, which is often less sustainable. Inequities also mean that vulnerable populations face higher risks of food insecurity.

The Triple Challenge encompasses three interconnected and urgent issues that threaten global food security: environmental degradation, climate change, and economic inequalities. Ecological degradation, caused by activities such as deforestation, soil erosion, water pollution, and biodiversity loss, diminishes vital natural resources necessary for resilient agriculture, reducing soil fertility and water quality. Climate change, driven by human-induced greenhouse gas emissions, leads to unpredictable weather patterns, droughts, floods, and rising temperatures, which severely disrupt planting and harvesting cycles and jeopardise crop yields. At the same time, economic inequalities restrict access to land, resources, and opportunities for marginalised populations, making it difficult for smallholder farmers and vulnerable communities to adopt sustainable practices or invest in productivity-enhancing technologies. These three challenges are deeply intertwined; addressing them requires integrated solutions that promote environmental sustainability, climate resilience, and social equity to ensure sufficient, equitable food production for all.

1.4 A Vicious Cycle

The three challenges create a vicious cycle. Environmental degradation exacerbates the impacts of climate change, which in turn threatens food systems. Economic inequalities hinder efforts to adapt to these changes, leaving marginalised groups more vulnerable to food insecurity. These challenges are global, affecting both developed and developing nations differently, but they collectively threaten the sustainability of the global food production system. They form a complex web of challenges that undermine food security worldwide. Addressing these issues requires an integrated approach that promotes sustainability, justice, and resilience in global food systems.

Economic inequalities in India are a profound challenge, rooted in its history, social structure, and development. Despite rapid growth, disparities in income, wealth, and access to resources remain widespread. A small elite hold most of the wealth, while many rural populations face poverty, malnutrition, and limited access to education and healthcare. Unequal access to resources like education, healthcare, and sanitation further perpetuates social divides, mainly affecting marginalised groups such as castes, tribes, and women. Regional disparities are also prominent, with urban centres thriving while many rural areas lack basic infrastructure, fuelling migration and placing stress on cities. Addressing these inequalities is vital for sustainable and inclusive growth.

Economic inequalities in India have a significant impact on agriculture, creating disparities in access to resources, technology, and opportunities among farmers. Wealthier farmers and large landholders often have better access to quality seeds, fertilisers, irrigation, credit, and modern machinery, enabling higher productivity and income. In contrast, small and marginal farmers, especially in rural and impoverished regions, lack access to these resources, which limits their productivity and keeps them trapped in a cycle of poverty. Economic disparities also restrict smaller farmers' ability to adopt sustainable and resilient farming practices, making them more vulnerable to climate change and market fluctuations. Additionally, inequalities in financial institutions and markets hinder equitable land ownership and access to government support programs, exacerbating rural poverty and preventing inclusive growth in the agricultural sector. Addressing these inequalities is essential to increasing productivity, ensuring food security, and promoting sustainable farming practices across all regions.

Economic inequalities in India have contributed significantly to malnutrition by limiting access to adequate food, healthcare, and sanitation for marginalised populations. Low-income families, especially in rural areas, often cannot afford nutritious food or healthcare services, leading to undernutrition and deficiencies. Limited access to quality education and sanitation further exacerbates malnutrition, as poor households struggle with issues like clean drinking water and hygiene, which are vital for health. Vulnerable groups such as children, women, and marginalised communities face higher risks of stunted growth, anaemia, and other health problems due to these disparities. Overall, economic inequalities hinder equitable access to resources essential for good nutrition, perpetuating a cycle of health and social disadvantages.

The vicious cycle created by these interconnected challenges perpetuates a reinforcing loop of decline in global food security. Environmental degradation intensifies the impact of climate change, such as droughts and floods, which further threaten food production systems worldwide. Climate-induced disruptions disproportionately affect marginalised communities, especially those with limited resources to adapt, deepening social and economic inequalities. These inequalities, rooted in historical and structural disparities, restrict access to vital resources like land, credit, and technology for smallholder farmers and vulnerable populations. This lack of access hampers their ability to adopt sustainable and resilient farming practices, making them even more susceptible to climate shocks and market fluctuations. Consequently, reduced productivity and increased vulnerability perpetuate poverty, malnutrition, and social exclusion. This feedback loop highlights the urgent need for integrated solutions that address environmental sustainability, social justice, and economic inclusion simultaneously, thereby breaking the cycle and ensuring resilient and equitable global food systems.

1.5 A New Set of Skills

Traditional leadership, focused mainly on increasing production and efficiency, must evolve into a more holistic approach that emphasises sustainability, environmental stewardship, and social responsibility. Future agricultural leaders must adopt innovative practices, integrate technology, and prioritise resource conservation to build resilient food systems. They also need to promote inclusive policies that bridge inequalities, empower marginalised communities, and ensure equitable access to resources and opportunities. This shift is crucial for fostering a sustainable, resilient, and equitable agricultural sector capable of meeting global food demand while protecting the environment for future generations.

A leader driving a paradigm shift in agriculture must possess a diverse set of skills. Innovative thinking is essential, enabling them to adopt and promote sustainable, eco-friendly, and cutting-edge farming practices. They require a profound understanding of environmental issues, including climate change and resource conservation, to make informed decisions that support sustainability. Familiarity with modern technology, such as digital tools, precision agriculture, and data-driven decision-making, is crucial for optimising productivity and resource use. An effective agricultural leader should also have inclusive leadership qualities, fostering collaboration and empowering marginalised communities to promote equitable resource distribution.

Strategic planning skills are vital for developing long-term, sustainable policies and practices that can withstand future challenges. Strong communication skills are essential for advocating sustainable practices, influencing stakeholders, and building consensus. Additionally, adaptability is key, enabling leaders to respond to changing environmental, social, and economic conditions. An entrepreneurial mindset facilitates innovation and the leveraging of private-public partnerships to promote sustainable development. Above all, ethical integrity and a commitment to community welfare and environmental stewardship underpin all these skills, ensuring that the leader's vision benefits both society and the planet.

Grooming leaders capable of driving a paradigm shift in agriculture requires a comprehensive approach that combines education, practical experience, and value-based training. Education programs should integrate technical knowledge with sustainable agriculture practices, climate resilience, and environmental conservation. Hands-on training through fieldwork, internships, and exposure to innovative farming methods can enhance practical skills. Leadership development programs must also focus on fostering critical thinking, ethical decision-making, and inclusive leadership qualities. Mentorship by experienced leaders, participation in policy discussions, and involvement in community projects can build confidence and social awareness. Encouraging interdisciplinary learning—combining agriculture, technology, environmental science, and social sciences—will prepare leaders to address complex challenges holistically. Furthermore, promoting a culture of innovation, entrepreneurship, and continuous learning is vital to ensure that future leaders stay adaptable and forward-thinking. By investing in these multifaceted development strategies, society can groom a new generation of agricultural leaders equipped to transform the sector sustainably and inclusively.

1.6 Interdisciplinary Learning and Innovation

Grooming transformative agricultural leaders demands more than academic excellence—it requires an ecosystem where learning, practice, and innovation flow seamlessly together. Universities and research institutions must not remain ivory towers; they need to extend their classrooms into farmers' fields, cooperatives, and rural enterprises. Field immersion programs, live demonstrations, and participatory research can help young leaders grasp the realities of farming beyond theory. Exposure to cutting-edge technologies, such as AI-driven precision farming, climate-resilient crop varieties, and renewable energy solutions, will enable them to bridge tradition with modern science. A holistic curriculum that balances ecological sensitivity with technological foresight will ensure that leaders are not just experts in their domain, but also visionaries who appreciate the interconnectedness of soil, water, climate, and human livelihoods.

At the same time, agricultural leadership cannot thrive without a clear understanding of market realities and socio-economic contexts. Leaders must be trained to navigate supply chains, understand consumer behaviour, and design solutions that are both environmentally sound and economically feasible. Collaborations with agribusinesses, policymakers, and farmer-producer organisations can foster this awareness, ensuring that innovation is not only practical but also scalable and inclusive. Creating forums for dialogue between farmers, scientists, and industry players encourages the co-creation of knowledge, while mentorship networks and peer learning platforms instil confidence and resilience in emerging leaders. Such an integrated and dynamic environment nurtures agricultural leaders who can synthesise science, practice, and commerce into a unified vision of sustainability, driving transformation that uplifts farmers, strengthens communities, and secures the future of food.

Lifelong employment for leaders driving a paradigm shift in agriculture is essential to ensure continuity, stability, and deep commitment to sustainable transformation. Given the complexity and long-term nature of environmental, social, and technological changes, these leaders must have the security and motivation to stay dedicated to their mission over extended periods. Lifelong employment encourages ongoing innovation, strategic planning, and adaptation without the distraction of changing roles or job insecurity. It also fosters a sense of ownership and responsibility, motivating leaders to continually update their skills, pursue best practices, and build institutional knowledge. Additionally, such stability enables them to develop long-term relationships with communities, policymakers, and industry stakeholders, which are essential for implementing sustainable strategies. Overall, lifelong employment provides the stability necessary for these leaders to effectively guide agriculture into a more sustainable, resilient, and inclusive future.

The envisioned future for agricultural leadership is one where a dynamic, collaborative ecosystem nurtures innovative, socially responsible, and technically proficient leaders committed to transformative change. This environment seamlessly integrates academia, practical fieldwork, and market realities to foster interdisciplinary learning, driving sustainable agricultural practices that are both environmentally sound and economically viable. Universities and research institutions serve as the knowledge hubs, emphasising hands-on training, exposure to emerging technologies, and real-world challenges. At the same time, active engagement with farming communities and industry stakeholders ensures that solutions are grounded in practical relevance and market needs.

Such a nurturing ecosystem promotes continuous dialogue, exchange of ideas, and adaptive learning, empowering leaders to address complex agricultural issues holistically. Ultimately, this integrated, ongoing support system will cultivate a new generation of visionary leaders who can steer agriculture towards a future of sustainability, resilience, and inclusivity—transforming the sector into a powerful driver of ecological health and social well-being for generations to come.

1.7 A Green Wealth Era

Agricultural and environmental leadership would drive innovation in eco-friendly farming, renewable resources, and technology, fostering an economy that values natural capital and ecological balance. People can thrive by engaging in productive, sustainable livelihoods that promote biodiversity, conserve resources, and reduce pollution. This ecosystem encourages equitable opportunity, empowering marginalised communities and rural populations to participate in wealth creation through sustainable agriculture, green technologies, and eco-tourism. Prosperity in the Green Wealth Era is measured not only by financial gains but also by improved health, social well-being, and environmental integrity. Such prosperity is rooted in sustainability, equitable growth, and environmental health, moving beyond traditional measures of wealth based solely on financial assets and defined by the well-being of both people and the planet, emphasising the importance of healthy ecosystems, clean resources, and social equity.

Economic progress drives innovation, development, and improved living standards, but it should do so in a way that does not compromise the health of our planet or marginalise vulnerable populations. Environmental sustainability ensures that natural resources are preserved and ecosystems are protected, providing the foundation for long-term economic stability and human well-being. Simultaneously, social justice guarantees equitable access to resources, opportunities, and benefits, ensuring that marginalised communities are not left behind and that development is inclusive and fair. Achieving this harmony requires integrating sustainable practices into economic policies, promoting green technologies, and ensuring social equity through fair distribution of resources and participatory decision-making. This balanced approach fosters a world where growth does not come at the expense of the environment or social cohesion, enabling communities to thrive sustainably and fairly, and securing a healthier, more equitable future for all.

When eco-friendly practices, such as regenerative agriculture, renewable energy, and resource conservation, become the foundation for economic development, they will generate new green industries and job opportunities. Education and skills development, with a focus on environmental stewardship and innovation, would empower individuals to make meaningful contributions to a resilient economy. Health and social well-being would flourish as pollution decreases, waste is minimised, and ecosystems are restored, leading to improved quality of life. Ultimately, prosperity in the Green Wealth Era reflects a balanced harmony between economic progress, environmental sustainability, and social justice, ensuring that growth benefits current and future generations while preserving the planet's ecological integrity.

If a little help is given to those farmers in 100 districts under the PM Dhan Dhanya Krishi Yojana, then the farmers there will also be on par with other farmers of India.

Narendra Modi

India@100 envisions a future where agricultural and environmental leadership catalyses a transformative Green Wealth Era, marked by innovation in eco-friendly farming, renewable resources, and cutting-edge technologies. In India, natural capital and ecological balance are crucial to economic prosperity, as they support livelihoods that are rooted in biodiversity conservation, resource efficiency, and pollution reduction. The economy thrives through sustainable industries such as regenerative agriculture, green energy, and eco-tourism, generating abundant green jobs and empowering marginalised rural communities to participate meaningfully in wealth creation. Through integrating sustainability into economic policies and fostering participatory decision-making, India@100 embodies a harmonious balance of growth, ecological integrity, and social inclusion.

Summary

This chapter emphasises a transformative shift from traditional food production to a sustainable approach that prioritises environmental health and livelihoods. It explores the interconnected challenges of environmental degradation, climate change, and economic inequalities, which perpetuate a vicious cycle of farmer poverty, malnutrition, and societal disparity. To address these issues, the chapter advocates for the development of new leadership skills rooted in interdisciplinary knowledge and cooperative spirit. These emerging Green Leaders are poised to spearhead a Green Economy, promoting sustainability, equity, and resilience, thereby fostering a more sustainable and just society.

Academic Exercise

1. What is the hallmark of 21st-century Century Agriculture?
2. Define Triple Challenge to global food security.
3. Describe the vicious cycle perpetuating poverty in the developing world.
4. Enumerate the new skill set of a transformative agriculture leader.
5. Outline vision of India@100 Economy based on a sustainable and just society.

2

Understanding Smallholders

> ***Smallholders may farm on small plots, but they carry the weight of the world's food security on their shoulders.***

2.1 The Heart of Global Food Security

Approximately 85-90% of Indian and African farmers are smallholders, defined as those owning less than 2 hectares (about 5 acres) of land. In China, 80-85% of Chinese farmers own less than 0.5 hectares of land. In South America, a smallholder owns or manages typically less than 5 hectares. These smallholders are indeed central to global food security. They contribute significantly to local and national food supplies, especially in developing countries, by practising sustainable farming methods and providing a diverse range of crops. Supporting smallholders through access to resources, markets, technology, and credit can enhance productivity, improve livelihoods, and strengthen food systems worldwide. Their role is vital in ensuring a resilient and inclusive global food network.

Smallholders are often referred to as the "heart of food security" because of their vital role in ensuring a stable, diverse, and accessible food supply. They produce a significant share of the world's food, especially in rural and developing regions, and contribute to local dietary diversity through the cultivation of a variety of crops. Their farming practices support millions of families, helping to reduce poverty and hunger while promoting resilience against environmental changes and climate variability. Additionally, smallholders help preserve biodiversity by growing traditional and indigenous crops, which are crucial for maintaining genetic diversity in agriculture and assume a central role in a sustainable agriculture system.

Smallholder farming sustains millions of families by providing them with the primary source of income, food, and livelihood in many rural areas around the world. These small-scale farmers often operate on limited land but diversify their crops and livestock to meet local needs and ensure food security for their households and communities. By producing a substantial portion of staple foods, smallholders help reduce hunger and improve nutritional outcomes. Additionally, their farming activities contribute to local economies, creating jobs and supporting rural development. Smallholders often sell surplus produce in local markets, generating income that can be reinvested into their farms or used to meet other basic needs. Overall, smallholder farming is a cornerstone of rural resilience, ensuring millions of families have the means to thrive and break the cycle of poverty and hunger.

Smallholders possess a unique advantage in their ability to adapt swiftly to changing environmental conditions, such as climate variability, droughts, or floods, due to their close connection to their land and flexible farming practices. Because they manage smaller plots of land, they can experiment with diverse cropping systems, adopt new techniques, and implement innovative solutions more rapidly than larger agricultural operations. This agility allows them to respond to adverse weather patterns, pests, or resource scarcity effectively, thereby reducing their vulnerability and maintaining productivity. Furthermore, smallholders often practice sustainable farming methods that preserve soil health, conserve water, and reduce chemical inputs, which contribute to long-term environmental stability. Their intimate knowledge of local ecosystems, combined with community-based approaches, encourages the use of eco-friendly techniques such as crop rotation, agroforestry, and organic farming. Promoting and supporting these sustainable practices not only helps smallholders adapt to climate change but also enhances biodiversity, improves soil fertility, and reduces the overall environmental footprint of agriculture, ensuring resilient farming systems for future generations.

India is renowned for its rich diversity of traditional and indigenous crops that are vital for maintaining genetic biodiversity. These include a variety of millets such as finger millet (ragi), pearl millet (bajra), foxtail millet, barnyard millet, and little millet, which are highly nutritious and resilient to harsh environmental conditions. Indigenous pulses like Bengal gram (chickpeas), black gram (urad), and horse gram also play a crucial role in traditional diets and agricultural sustainability. The country is also home to native oilseeds such as mustard, neem, and wild varieties of sesame, as well as unique fruits like aonla (Indian gooseberry), bael, and custard apple. Traditional vegetables, including indigenous gourds, bitter gourd, and nutrient-rich leafy greens like amaranth and drumstick, further enrich the biodiversity. Additionally, ancient varieties of rice and wheat, adapted to diverse ecological zones, contribute to India's agricultural resilience. All these are smallholder enterprises.

2.2 Profile of smallholder farmers in India and the Global South

Smallholder farmers in India have mostly relied on traditional farming methods to cultivate staple crops such as rice, wheat, millet, and pulses. These farmers often depended on family or community labour and have faced challenges like limited access to quality seeds, fertilisers, and irrigation. They have also been vulnerable to market fluctuations and climatic risks such as droughts and floods, which significantly impacted their productivity and income.

In recent years, the Indian government has implemented various schemes and initiatives to support smallholder farmers and improve their livelihoods. Programs such as the Pradhan Mantri Krishi Sinchayee Yojana (PMKSY) aim to enhance irrigation infrastructure, helping farmers reduce dependency on monsoon rains and increase crop productivity. The Pradhan Mantri Fasal Bima Yojana (PMFBY) provides affordable crop insurance coverage to protect farmers from crop failure due to natural calamities and pest outbreaks, offering financial security and encouraging risk-taking. Additionally, schemes like the Soil Health Card and Neem Coated Urea promote sustainable farming practices by improving soil quality and reducing environmental impact.

Furthermore, the government has introduced schemes to improve market access and income for smallholders. The e-NAM (National Agriculture Market) platform integrates farmers with a broader network of traders and buyers, enabling better price discovery and transparency. Several subsidy programs provide affordable access to quality seeds, fertilisers, and equipment, while training and extension services aim to improve farming techniques. These efforts collectively seek to boost productivity, increase rural incomes, and promote sustainable agriculture among smallholder farmers across India.

The Kisan Rail initiative utilises India's railway network to quickly transport perishable farm produce like fruits and vegetables from surplus regions to consumption centres, connecting farmers to larger markets and reducing wastage. The Kisan Vikas Portal (also known as the e-NAM platform) provides a pan-India, online trading platform for agricultural products, connecting farmers to buyers and facilitating transparent price discovery, thereby promoting better market access and export opportunities. These initiatives, along with the Krishi Udan Scheme for air freight, aim to strengthen the agricultural supply chain, boost farmer incomes, and improve overall access to markets for farm produce.

India is often considered a model for the Global South in agriculture due to its significant progress in certain areas of agricultural development and systemic reforms. India has achieved self-sufficiency in food production, especially in staple crops like rice and wheat, through the Green Revolution of the 1960s and 1970s, which introduced high-yielding variety seeds, improved irrigation, and modern farming techniques. This transformation enabled India to become one of the world's largest producers of food grains, thereby reducing its dependence on imports and enhancing food security.

While Indian farmers were not necessarily the very first in the world to adopt hybrid seeds, India was among the first countries in Asia to widely embrace them as part of a concerted effort to boost agricultural productivity. The successful adoption of hybrid seeds, along with related technologies like chemical fertilisers and improved irrigation, played a crucial role in transforming Indian agriculture during that period. This pioneering adoption served as a model for other developing countries seeking to increase crop yields through modern seed technology. Dr M.S. Swaminathan wrote:

> Brimming with enthusiasm, hard-working, skilled and determined, the Punjab farmer has been the backbone of the revolution. Revolutions are usually associated with the young, but in this revolution, age has been no obstacle to participation. Farmers, young and old, educated and uneducated, have easily taken to the new agronomy. It has been heartwarming to see young college graduates, retired officials, former army personnel, illiterate peasants, and small farmers queuing up to receive the new seeds. At least in Punjab, the divorce between intellect and labour, which has been the bane of our agriculture, is vanishing.[1]

However, while India offers important lessons in scaling agricultural productivity and implementing large-scale policy reforms, it also highlights challenges that many developing countries face, such as disparities in income, access to resources, and sustainability issues. The country's ongoing efforts to improve rural infrastructure, expand access to technology, and promote sustainable practices serve as valuable models for other nations. Nonetheless, the complexities of smallholder farmer livelihoods, environmental concerns, and market access require context-specific solutions. Overall, India's experience provides both inspiration and

---

[1] https://www.ias.ac.in/article/fulltext/reso/019/02/0109-0115. Last accessed on August 15, 2025.

caution for the Global South in navigating agricultural growth and rural development.

2.3 Socioeconomic realities and constraints

Smallholders often face infrastructural challenges such as poor rural roads, limited access to markets, and inadequate storage facilities, which hinder their ability to sell produce at fair prices. They also encounter social constraints, including a lack of education, limited awareness of new farming techniques, and restricted access to government schemes or extension services. Environmental issues, such as recurring droughts, floods, and land degradation, further worsen their conditions. These socioeconomic barriers collectively restrict smallholder farmers from maximising their agricultural potential and improving their overall standard of living.

Farmers in North Bihar, for example, face a recurring and severe crisis due to annual flooding, which ravages their agricultural land and livelihoods. The region is highly vulnerable to seasonal floods caused by overflowing rivers such as the Ganges, Gandak, and Kosi, often exacerbated in the absence of a flood management infrastructure. The Kosi River is even called the "Sorrow of Bihar". As floods submerge farmland, farmers lose their crops, livestock, and sources of income, plunging them into long-term economic hardship. The destruction of crops not only impacts the current harvest but often diminishes land fertility, affecting future yields as well. The recurring nature of the floods also hampers efforts to adopt modern agricultural practices or invest in resilient crops.

India has several regions that are regularly affected by drought due to low and unpredictable rainfall, limited water resources, and semi-arid or arid conditions. Rajasthan is the most prominent drought-prone state, especially in its western parts like Jaisalmer,

Barmer, and Jalore, where the desert landscape and scarce rainfall make agriculture and water availability extremely challenging. Similarly, parts of Maharashtra, such as Marathwada and Vidarbha, frequently face drought conditions, severely impacting farmers and local communities. In Odisha, western districts often experience droughts that threaten crop production and drinking water supplies. At the same time, regions in Gujarat, especially Kutch and the southern parts, are also vulnerable due to uneven and scant rainfall.

Other drought-prone areas include parts of Punjab and Haryana, which face groundwater depletion and periodic droughts, especially in western regions. Northern Karnataka, particularly in districts like Bellary and Bagalkot, struggles with chronic water scarcity that hampers agriculture. Telangana and Andhra Pradesh, especially the Rayalaseema region, are also prone to recurrent droughts, affecting local livelihoods. These areas share common challenges such as inadequate water management infrastructure, over-reliance on groundwater, and erratic monsoon patterns, making drought a persistent threat with serious socio-economic impacts on farming communities and local economies.

The three main vows of smallholders in India often emphasise the importance of sustainable farming practices. These farmers are committed to environmentally friendly methods that conserve natural resources, maintain soil health, and minimise the use of harmful chemicals. By adopting such practices, smallholders aim to ensure the long-term productivity of their land and contribute to ecological balance. This commitment reflects their recognition of the crucial role that sustainable agriculture plays in securing their livelihoods and protecting the environment for future generations.

In addition, smallholders in India prioritise crop self-sufficiency and community welfare as key objectives. They prioritise growing enough food for their own consumption, reducing dependence on external markets and ensuring food security for their families. Moreover, they often emphasise the importance of cooperation within their communities, promoting collective efforts to improve living standards and resist exploitative practices. These vows collectively foster resilience, social cohesion, and sustainable development among smallholder farmers in India.

2.4 The power dynamics in supply chains and market inequities

Power dynamics in supply chains and market inequities significantly impact Indian smallholders, often placing them at a disadvantage in the agricultural sector. Large corporations, middlemen, and export companies typically hold disproportionate power, controlling pricing, access to markets, and input supplies. Smallholders often lack bargaining power, which leaves them vulnerable to exploitation, low prices for their produce, and unfair contractual arrangements. As a result, they receive only a fraction of the consumer prices, while intermediaries and big players capture most of the profits, perpetuating income inequalities and economic dependency among small farmers.

Market inequities further exacerbate the struggles faced by Indian smallholders. Limited access to formal credit, modern technology, and market information restricts their ability to negotiate favourable deals and adopt innovative farming practices. Additionally, infrastructural challenges such as poor transportation, storage facilities, and market linkages diminish their bargaining strength and lead to post-harvest losses. These systemic inequalities hinder smallholders' growth prospects and push many into cycles of poverty, making it difficult for them to improve their livelihoods and achieve greater market participation.

> ***I am one of those who firmly believe that our cities thrive at the expense of our villages; that our industries exploit agriculture.***
>
> ***Verghese Kurien***

It is estimated that farmers in India receive barely one-fifth of what consumers pay for agricultural produce. The remaining 80% is absorbed along the supply chain—by commission agents, wholesalers, retailers, transporters, and storage operators—leaving cultivators with only a fraction of the value their hard work creates. This wide disparity is not merely an economic statistic; it reflects systemic inequities that have long burdened the agricultural sector.

The root causes are many. A dense network of intermediaries takes successive cuts, often leaving farmers powerless in negotiations. Logistical challenges—ranging from inadequate cold storage to high transportation costs—add layers of expense that further shrink their margins. In many cases, perishable produce forces farmers into distress sales, making them more vulnerable to exploitation.

This imbalance highlights the urgent need for supply chain reforms and the empowerment of farmers. Strengthening farmer-producer organisations (FPOs), cooperatives, and digital marketplaces can help cultivators bypass exploitative middlemen. Establishing robust direct-to-consumer models, such as farmers' markets and e-commerce platforms, can ensure fairer returns. Policy interventions that reduce transaction costs, improve infrastructure, and promote transparency in pricing will be critical to restoring balance. Ultimately, the goal must be to transform farmers from price-takers to price-makers, ensuring that the sweat and skill invested in the soil are rewarded with dignity and justice. Only then can agriculture become not just a means of survival, but a sustainable engine of prosperity for millions who feed the nation.

This unjust system is perpetuated through a combination of structural, institutional, and market-based factors. One key element is the dominance of middlemen and commission agents in rural markets, who act as gatekeepers and control the flow of produce from farmers to consumers. These intermediaries often have better access to market information, credit, and transportation, which allows them to exploit farmers by offering low prices for their produce. The lack of direct access to markets for farmers, coupled with limited bargaining power, sustains this unequal distribution of profits.

Institutional gaps also play a significant role in perpetuating the system. Weak infrastructure, inadequate market regulations, and limited enforcement of fair-trade practices hinder farmers' ability to negotiate better prices or sell directly to consumers. Additionally, the lack of transparency and bargaining mechanisms leaves farmers vulnerable to exploitation. Overall, a combination of entrenched power structures, infrastructural deficiencies, and market inefficiencies continues to sustain this unequal distribution of the value chain, preventing farmers from receiving their fair share.

The success of farmer cooperatives in India has been evident in several regions where they have played a crucial role in empowering farmers, ensuring fair prices, and improving access to inputs and markets. Successful cooperatives such as dairy, sugar, and cotton cooperatives have helped farmers achieve economies of scale, reduce exploitation by middlemen, and increase their bargaining power. These cooperatives have also facilitated access to credit, technology, and infrastructure, contributing to rural development and improved livelihoods. For example, the success of the Amul dairy cooperative revolutionised milk production in India, making dairy farming profitable for millions of small farmers.

However, many farmer cooperatives have also faced significant failures due to various challenges. These include weak governance, political interference, corruption, and mismanagement, which often lead to financial instability and loss of farmers' trust. In some cases, cooperatives have been manipulated for political gain or personal enrichment of leaders, rather than for the welfare of farmers. Overall, while some cooperatives have successfully transformed rural livelihoods, many have struggled or failed due to structural weaknesses and governance issues.

2.5 Impact of corporate interests and political barriers

A "money landing lobby" or financial interests stands like a shadow over the woes of small farmers in India. In 2018, the Yavatmal district in Maharashtra experienced a series of farmer suicides, which drew significant media attention and public concern. Many small farmers in the region were grappling with mounting debts, crop failures, and poor market prices, which severely impacted their livelihoods. The region, which is primarily dependent on agriculture, faced adverse weather conditions, including drought-like situations, that further exacerbated the farmers' economic plight.

Over the fields of small farmers looms a money-lending lobby, turning policies into profits for the powerful while pushing cultivators deeper into debt and inequity.

This wave of suicides underscored the deep-rooted issues of indebtedness, inadequate access to credit, and lack of effective government support for distressed farmers. The tragedy prompted debates on the need for comprehensive agricultural reforms, debt relief measures, and better mental health and social support systems for rural farmers. The Yavatmal incidents became a symbol of the ongoing agrarian crisis affecting many parts of India, highlighting the urgent need for policy intervention to address farmers' distress.

A lot has been done, and yet the role of commercial lenders, private financiers, and lobbying groups in creating debt traps for small farmers, deeper market inequities, and limited access to fair credit and support systems is undeniable. While factors such as market inefficiencies, lack of infrastructure, and climate change are also major reasons for farmers' distress, the influence of powerful financial and corporate lobbies is often discussed as a contributing factor that exacerbates existing vulnerabilities for smallholders.

Looking ahead, there is hope that the lessons learned from the Yavatmal incidents will catalyse meaningful change and sustainable solutions for Indian agriculture. Increased awareness has sparked government and community efforts to implement targeted reforms, including better access to credit, crop insurance, and climate-resilient farming techniques. Initiatives focusing on mental health support and social safety nets for farmers are gaining momentum, promising to create a more compassionate and inclusive rural support system. Moreover, ongoing dialogue among policymakers, agricultural experts, and farmers is paving the way for innovative strategies that can help mitigate risks, stabilise incomes, and ensure the well-being of farming communities.

Ultimately, the redemption of Indian agriculture lies not in subsidies or slogans, but in restoring dignity to those who till the land. The nexus of money, power, and policy must give way to a new covenant of trust between the farmer and the nation. True reform must rise from the soil upward — not as an act of charity, but as a moral awakening — where prosperity is shared and justice is rooted in compassion. When the farmer stands unburdened by debt and despair, sowing seeds not of survival but of hope, India will have reclaimed the truest meaning of freedom.

Summary

This chapter emphasises the vital role of smallholder farmers in ensuring global food security, highlighting their central position in feeding local and international populations. It provides a detailed profile of smallholders, illustrating their diverse backgrounds, farming practices, and economic conditions. The chapter explores the socio-economic realities they face, including limited resources, access to markets, and financial constraints, which often hinder their development. It also examines the complex power dynamics within supply chains and market systems, revealing how inequalities and corporate interests can marginalise smallholders. Furthermore, the chapter discusses political barriers and policy gaps that exacerbate their vulnerabilities. Overall, it underscores the importance of re-envisioning leadership to advocate for more equitable, inclusive, and sustainable support systems that empower smallholders and ensure their vital contribution to food systems worldwide.

Academic Exercise

1. How are smallholders at the heart of global food security?
2. Profile an Indian smallholder.
3. How did the pioneering adoption of the Green Revolution by Indian smallholders create a model for other developing countries?
4. Enumerate the non-farming factors affecting smallholders.
5. Emphasise the importance of long-term resilience for smallholder farmers.

3

Beyond Science

The leaders of tomorrow's agriculture will not stand in silos of science, but at the crossroads—where genetics meets ecology, economics meets policy, and innovation meets responsibility.

3.1 Interdisciplinary Future Agriculture

Envisioning the future of interdisciplinary agriculture involves the seamless integration of multiple fields, including technology, biology, environmental science, economics, and social sciences. This holistic approach aims to revolutionise traditional farming practices by making them more sustainable, efficient, and resilient to global challenges. Advances in precision agriculture, including the use of GPS, IoT sensors, and drones, allow farmers to optimise resource use such as water, fertilisers, and pesticides. This not only increases crop yields but also significantly reduces environmental impact. Additionally, biotechnology plays a crucial role in developing genetically modified crops and livestock that are resistant to pests, diseases, and climate stresses, thereby enhancing food security and crop resilience.

The future of agriculture also emphasises environmental sustainability by promoting practices such as conservation agriculture, which focuses on maintaining soil health through cover cropping, crop rotation, and minimal tillage. Resource management is increasingly relying on renewable energy sources, such as solar and wind power, to reduce carbon footprints. The concept of a circular economy is gaining importance, with waste recycling, composting, and biogas production turning farming waste into valuable resources, further closing the loop in agricultural systems.

Biological and ecological sciences are fundamental to designing systems that work harmoniously with natural ecosystems. Agroecology, for example, promotes farming practices that leverage biodiversity and natural pest control, reducing the reliance on chemical inputs. Enhancing soil and water science is also critical, with efforts focused on maintaining long-term fertility and efficient water use, which are vital in the face of climate change. These biological sciences foster a deeper understanding of ecosystems, guiding the development of farming methods that are both productive and environmentally friendly.

On the social and economic front, future agriculture is shaped by community engagement, policy development, and education. Encouraging local participation ensures that farming practices are culturally appropriate and socially accepted. Developing supportive policies and providing training programs will be essential to foster innovation and help farmers adopt sustainable technologies. Digital platforms and transparent supply chains can enhance market access, promote fair trade, and increase consumer awareness, ultimately creating a more equitable and responsive food system.

Addressing the social dimensions of agriculture ensures that technological advances benefit all stakeholders, fostering social equity and resilience.

3.2 An Integrative Approach

Envisioning the integration of seed genetics, agronomy, soil science, ecology, and social sciences presents a comprehensive approach to transforming agriculture into a more sustainable and resilient system. Seed genetics serves as the foundation for developing crops that are not only high-yielding but also resistant to pests, diseases, and climate stresses. These improved seeds can be tailored for specific environments, enabling farmers to adapt to changing conditions and increase overall productivity. Complementing this, agronomy provides the science of crop management—optimising planting schedules, crop rotation, fertilisation, and irrigation to maximise resource efficiency and reduce environmental footprints.

Soil science plays a critical role in maintaining and enhancing soil health, which is essential for sustainable crop production. By studying soil properties and nutrient cycles, soil scientists can recommend practices that enhance soil structure, improve fertility, and promote carbon sequestration, thereby mitigating soil degradation. Ecology integrates into this system by encouraging farming practices that support biodiversity, natural pest regulation, pollination, and other ecosystem services. This ecological perspective aims to design farming systems that coexist harmoniously with natural habitats, reducing dependence on chemical inputs and increasing resilience against environmental stresses.

Meanwhile, social sciences ensure that these scientific and technological advancements are effectively adopted and accepted by farming communities. Understanding farmers' social, economic, and cultural contexts helps tailor solutions that are accessible, equitable, and culturally appropriate. It also addresses issues related to education, policy, and market access, fostering an environment that enables innovation dissemination and sustainable development.

Together, these disciplines foster a systems-based approach that not only increases agricultural productivity but also preserves natural resources, enhances environmental stability, and promotes social equity. The integration of seed genetics, agronomy, soil science, ecology, and social sciences creates a dynamic and adaptive agricultural model capable of meeting current and future challenges, ensuring food security while maintaining the planet's health, which is indeed a central theme of 21st-century farming.

Overall, this theme underscores the importance of developing resilient, efficient, and environmentally friendly agricultural systems that can nourish a growing global population without compromising the health of the planet for future generations.

3.3 Economics, law, and policy in sustainable agriculture

Economics, law, and policy play crucial roles in shaping sustainable agriculture by creating an enabling environment for environmentally friendly practices that are economically viable and socially equitable. Economic incentives such as subsidies, carbon credits, and market-based mechanisms encourage farmers to adopt sustainable technologies and practices that reduce environmental impact while maintaining profitability. Proper valuation of ecosystem services and natural resource management can also influence decision-making, ensuring that ecological benefits are recognised and preserved.

Legal frameworks and regulations are essential for enforcing environmental standards, protecting water and soil resources, and promoting conservation efforts. Laws that regulate chemical use, land tenure, and biosecurity help safeguard ecological health and ensure sustainable land management. Moreover, policies that support research, extension services, and education help disseminate sustainable practices and technological innovations among farmers.

In India, the legal frameworks and regulations for enforcing environmental standards and protecting water and soil resources are well established, but face challenges related to implementation and enforcement. The primary legislation includes the Environment Protection Act (1986), which provides a comprehensive framework for environmental regulation, along with specific laws such as the Water (Prevention and Control of Pollution) Act (1974), the Air (Prevention and Control of Pollution) Act (1981), and the Forest Conservation Act (1980).

The Central and State Pollution Control Boards (CPCB and SPCBs) are responsible for monitoring compliance, issuing permits, and taking enforcement actions. The National Green Tribunal (NGT), established in 2010, serves as a dedicated judicial body to resolve environmental disputes, enforce legal standards, and ensure compliance with environmental laws.

Despite the robust legal framework, enforcement challenges persist. These include inadequate monitoring, lack of technical capacity at local levels, and pollution from industrial, agricultural, and urban sources. Moreover, land use violations, illegal mining, and unregulated industrial activities often undermine environmental regulations, especially in rural and semi-urban regions.

Recent initiatives such as the National Clean Water Mission and the Swachh Bharat Abhiyan aim to improve water quality, sanitation, and waste management. However, balancing development needs with environmental protection remains complex. Strengthening enforcement mechanisms, increasing awareness, and integrating sustainable practices into agricultural and industrial development are ongoing priorities to safeguard water and soil resources in India better.

Effective policy-making involves integrating scientific insights with economic and legal considerations to develop comprehensive strategies that promote resilience, conserve natural resources, and improve livelihoods. International agreements and national policies also play a crucial role in addressing global challenges, such as climate change and food security, by fostering cooperation and shared responsibility. Ultimately, aligning economics, law, and policy is essential for building a sustainable agricultural future that balances productivity, ecological health, and social well-being.

3.4 Case studies of successful interdisciplinary approaches

India offers several compelling case studies that demonstrate successful interdisciplinary approaches to sustainable agriculture, integrating science, technology, policy, and community involvement. Starting with the Green Revolution in the 1960s, India embarked on a transformative journey towards self-sufficiency in food production. The Green Revolution introduced high-yielding varieties (HYVs) of wheat and rice, along with modern irrigation techniques, chemical fertilisers, and mechanisation. This revolution significantly increased crop yields, reducing dependence on food imports and averting famines. It also spurred the development of rural infrastructure, improved agricultural productivity, and fostered the growth of ancillary industries such as fertiliser, pesticide, and machinery manufacturing.

While the Green Revolution achieved remarkable success in boosting food grain production, it also brought challenges such as environmental degradation, groundwater depletion, soil fertility decline, and increased income disparities among farmers. Despite these issues, India never looked back after this initial burst of agricultural growth. The focus shifted toward sustainable practices, integrating science and technology with policy reforms, farmer training, and diversification efforts. Over the decades, India has continued to innovate in agriculture through various initiatives like climate-resilient crops, organic farming, and digital agriculture, all building on the foundational achievements of the Green Revolution. The revolution marked a critical turning point—propelling India from food scarcity to a major agricultural producer—and set the stage for ongoing efforts towards sustainable and inclusive growth in the sector. From many, three examples are cited here.

3.4.1 System of Rice Intensification (SRI)

Originating in Madagascar in the 1980s, the System of Rice Intensification (SRI) has gained widespread popularity globally, including in India, due to its effectiveness and sustainability benefits. SRI involves a set of specific farming practices that differ significantly from conventional rice cultivation methods. These include planting younger, single seedlings with wider spacing, maintaining soil aeration through methodical watering practices, and using organic manure instead of chemical fertilisers.

SRI combines insights from agronomy, ecology, and traditional knowledge to increase rice productivity while reducing water use and chemical inputs. The approach involves specific practices such as altered planting techniques, organic soil management, and increased plant spacing, which rely on ecological principles and community-led adaptation. SRI is practised in states like Tamil Nadu and Odisha. The approach emphasises soil health and ecosystem balance, encouraging benefits such as reduced water consumption—often by up to 30-50%—and lower pesticide use. Additionally, SRI boosts plant resilience and yields by promoting healthy root and plant development.

3.4.2 Rainwater Harvesting and Watershed Management

Another significant case is the Rainwater Harvesting and Watershed Management Projects in regions like Rajasthan and Maharashtra. These projects integrate soil science, water resource management, ecology, and local community participation. By promoting rainwater harvesting, groundwater recharge, and sustainable land use practices, they have helped restore degraded ecosystems, improve agricultural productivity, and ensure water availability for farmers. The success of these projects is rooted in integrating scientific research with policy frameworks and community engagement.

> *Water harvesting is not some ATM into*
> *which you can put a card and get water.*
>
> ***Anupam Mishra***

India has undertaken numerous large-scale Rainwater Harvesting and Watershed Management projects to address water scarcity, restore degraded ecosystems, and boost agricultural productivity. Rajasthan, one of the most arid states, has been a pioneer in this area since the 1980s. Its watershed management efforts focus on land treatment, reforestation, and rainwater harvesting techniques such as contour trenches, check dams, and percolation tanks. The Rajasthan Watershed Development Project successfully recharged groundwater tables, improved rural livelihoods, and curbed land degradation. Building on these successes, the government launched the Integrated Watershed Management Program (IWMP), which has been adopted nationwide to promote similar ecological and water conservation measures in semi-arid regions across India.

In Maharashtra, drought-prone districts like Marathwada and Vidarbha have seen significant benefits from watershed initiatives such as Jalkavanish Yojana (or more accurately, the Jalyukt Shivar Abhiyan) and the Paani Foundation's work, which create structures like farm ponds, check dams, and contour trenches to capture and retain rainwater and surface runoff, leading to better groundwater recharge and more stable water supply. Similarly, in Karnataka, projects in the Kolar and Chitradurga districts focus on hillside terracing, afforestation, and rainwater harvesting to restore natural water sources. These efforts have successfully increased groundwater levels, supported small-scale irrigation schemes, and improved rural livelihoods by revitalising ecosystems.

On a broader scale, the Ganga River Basin Management Plan exemplifies a comprehensive, integrated approach to watershed management. Key to this initiative is cooperation across multiple states to rejuvenate the Ganga River through pollution control, rainwater harvesting, and sustainable water management practices. The strategy emphasises river basin approaches that balance ecological needs with socio-economic development. Overall, these projects underscore the importance of interdisciplinary strategies that combine engineering, ecology, community participation, and government support. Their success demonstrates that integrated rainwater harvesting and watershed management are crucial for addressing water scarcity, restoring ecosystems, and enhancing rural resilience in India's diverse landscapes.

3.4.3 Agroforestry

The Agroforestry initiatives in states like Kerala and Madhya Pradesh highlight interdisciplinary success by blending biodiversity conservation, soil health, climate resilience, and social welfare. These programs incorporate ecological principles, farmer-led practices, and policy support to develop diversified cropping systems that promote ecological balance, increase income, and enhance resilience against climate-related shocks.

In Kerala, agroforestry practices are often rooted in traditional knowledge but have been modernised with scientific guidance to optimise land use. Farmers are encouraged to plant multipurpose trees such as coconut, banana, medicinal plants, and fruit trees alongside staple crops like rice, pepper, and spices. This diversified approach helps in creating microclimates, reducing soil erosion, and maintaining soil fertility through organic matter input. As a result, farmers benefit from multiple sources of income—timber, fruits, fodder, and medicinal products—while also improving ecological resilience against climate fluctuations.

Government policies and programs have played a supportive role in expanding agroforestry in Kerala. The State Forest Department and agricultural agencies promote agroforestry through training, demonstrations, and subsidy schemes for tree planting. Initiatives like the "Green Kerala" campaign aim to increase tree cover in both rural and urban areas, emphasising the importance of ecological sustainability and climate change mitigation.

Many farmers in Kerala have adopted agroforestry practices on their farms, resulting in tangible benefits. These include increased crop yields, improved soil and water conservation, and enhanced biodiversity. The integration of medicinal plants, especially in the hill regions, has also opened new avenues for livelihood diversification. Furthermore, agroforestry contributes to the state's efforts in carbon sequestration and environmental protection, aligning with broader climate resilience goals.

Agroforestry in Madhya Pradesh embodies a strategic approach to land management that integrates the cultivation of trees with agricultural crops to promote sustainability, biodiversity, and rural livelihoods. This region, characterised by diverse agro-climatic zones ranging from forested hills to arid plains, has adopted agroforestry practices to enhance productivity, conserve natural resources, and mitigate the impacts of climate change. Farmers typically integrate multipurpose trees, such as neem, acacia, babul, and Amla (Indian gooseberry), alongside food crops like wheat, soybeans, and pulses. The practice also helps improve soil fertility through nitrogen fixation and the addition of organic matter, while tree roots reduce soil erosion, especially in sloping terrains.

The government and non-governmental organisations actively promote agroforestry through various initiatives. Schemes like the National Mission for Sustainable Agriculture (NMSA) and state-level programs encourage farmers to adopt agroforestry by providing technical assistance, training, demonstration plots, and financial incentives. These efforts aim to increase tree cover, especially in degraded lands, thereby supporting ecological restoration and climate resilience.

3.5 Innovative Educational Strategies

Educational reforms for interdisciplinary agriculture focus on integrating diverse fields such as biotechnology, ecology, soil science, economics, and policy studies into agricultural education to meet the evolving demands of sustainable and innovative farming systems. Traditionally, agricultural education has been centred around purely technical skills, but modern challenges require a holistic approach that combines scientific knowledge with socio-economic considerations.

Reforms can include updating curricula to incorporate cross-disciplinary courses that cover environmental science, climate change mitigation, digital technologies, and social sciences alongside traditional agronomic studies. Encouraging project-based learning, field internships, and collaborations with research institutions fosters practical understanding and innovation. Establishing specialised departments or programs in agroecology, sustainable agriculture, and climate-resilient farming can prepare students to develop integrated solutions.

Agroecology is a comprehensive approach to agriculture that emphasises sustainability by integrating ecological principles into farming systems. It focuses on enhancing biodiversity, improving soil health, and promoting resource efficiency to create resilient and productive agricultural ecosystems. This approach encourages practices such as crop diversification, organic farming, composting, cover cropping, and the integration of crops and livestock to optimise nutrient cycling and reduce dependence on chemical inputs like synthetic fertilisers and pesticides. By fostering harmonious interactions among plants, animals, humans, and their environment, agroecology seeks to enhance ecosystem stability, increase ecological resilience, and support the livelihoods of local communities and farmers. It also promotes social equity by empowering farmers, encouraging fair practices, and strengthening local food systems. The benefits of agroecology include reducing environmental pollution, conserving water, improving soil fertility, and contributing to climate change mitigation.

Agroecology is not just farming—it is the art of working with nature, where biodiversity becomes resilience, soil becomes legacy, and farmers become custodians of both food and future.

Furthermore, promoting multi-disciplinary research and capacity building through workshops, seminars, and faculty exchanges broadens perspectives and enhances problem-solving skills. Introducing coursework on agribusiness management, policy formulation, and rural development ensures that future agricultural leaders are equipped to address socio-economic challenges and implement effective policies. Educational reforms also emphasise the importance of community engagement and participatory learning methods, enabling students to work directly with farmers and local communities. This fosters a deeper understanding of ground realities and encourages innovative, context-specific solutions.

The apprentice scheme in agriculture is a structured program designed to provide hands-on training and practical experience to aspiring farmers, agricultural students, or rural youth. This scheme aims to bridge the gap between theoretical knowledge and real-world farming practices, enabling participants to develop skills in modern agricultural techniques, technology adoption, and sustainable farming methods. Typically, the apprentice scheme involves pairing trainees with experienced farmers, agronomists, or agriculture experts who mentor them on the day-to-day operations of farm management, crop cultivation, irrigation, pest control, soil health management, and post-harvest technologies. The program often spans several months to a year, during which apprentices gain invaluable on-ground experience, learn innovative practices, and understand the economics of farming.

In India, Farmer Producer Organisations (FPOs) have gained prominence as a key strategy to empower farmers, especially in marginal and small holdings. They are typically registered as non-profit or cooperative societies under the Companies Act or Cooperative Societies Act and are supported by government schemes, banks, and development agencies such as NABARD (National Bank for Agriculture and Rural Development). The core activities of FPOs include bulk procurement of inputs like seeds, fertilisers, and pesticides, shared access to better technology and extension services, collective marketing of produce, and value addition.

FPOs also facilitate capacity building among farmers through training, skill development, and technical support, which enhances their ability to adopt sustainable and innovative farming practices. By operating collectively, farmers can negotiate better prices for their products, reduce transaction costs, and access credit and insurance facilities more effectively. Additionally, FPOs contribute to rural employment, ecosystem management, and social inclusion, empowering marginalised communities and women.

The success of FPOs largely depends on effective governance, proper financial management, and support from government policies. Many states have launched specific programs and subsidy schemes to promote the formation and strengthening of FPOs, recognising them as a vital model for inclusive agricultural growth. Overall, FPOs represent a promising approach to democratise agricultural markets, improve farmers' livelihoods, and promote sustainable rural development through collective effort and institutional strengthening.

Summary

This chapter emphasises the necessity of developing a new style of leadership to shape the future of sustainable agriculture. It highlights the interdisciplinary nature of modern agriculture, advocating for the integration of seed genetics, agronomy, soil science, ecology, and social sciences to create innovative and resilient farming systems. The chapter highlights the crucial roles of economics, law, and policy in promoting sustainable practices, safeguarding environmental protection, and fostering equitable development. Through case studies of successful interdisciplinary approaches, it showcases practical models of innovation and collaboration. The conclusion emphasises the urgent need for educational reforms and the implementation of robust apprenticeship schemes to attract young people into agriculture, fostering leadership that combines scientific knowledge, ecological awareness, and social responsibility for a sustainable agricultural future.

Academic Exercise

1. Enumerate the fields involved in interdisciplinary future agriculture.
2. What is a systems-based approach for agriculture?
3. Describe economic incentives for agriculture with examples.
4. Define Agroecology.
5. What are Farmer-Producer Organisations (FPOs)?

4

Rethinking Technology

The future of farming lies not in universal fixes, but in intelligent tools that listen to the land and respond to its unique voice.

4.1 From GMOs to Biological Fertilisers and AI

A rethink of technology in agriculture is necessary because crop production is at a crossroads, where traditional practices no longer suffice to meet global demands. Advances such as GMOs, biological fertilisers, and AI have demonstrated their potential to significantly improve crop yields, reduce resource use, and promote more sustainable farming methods. However, embracing these technologies also requires careful consideration of their long-term impacts, ethical implications, and social acceptance. A re-evaluation ensures that innovation aligns with environmental sustainability and food security goals, avoiding unintended consequences.

Moreover, the rapid pace of technological development calls for a reassessment of current policies, regulations, and practices to ensure they are conducive to responsible innovation. While these tools offer promising solutions, such as resilience to climate change and reduced chemical inputs, they also pose new challenges, including regulatory gaps and potential ecological risks. Rethinking technology in agriculture involves balancing the benefits of modernisation with the need for safety, ethics, and social equity. However, the question is who would do it and whether there is a political consensus on the matter.

This shift also encourages a more inclusive approach, involving farmers, scientists, policymakers, and communities to shape technology adoption in ways that are contextually appropriate and socially acceptable. Ultimately, a comprehensive re-examination of technological strategies is essential to develop a resilient, sustainable, and equitable agricultural system capable of feeding a growing world population while preserving the environment for future generations. This is about creating awareness through public discourse and media. As this is an interdisciplinary matter, scientists tend to evade speaking beyond their domain expertise, and it is left to journalists.

AI has the potential to bridge the gap between science and social welfare by enabling more informed, data-driven decision-making that considers both technological advancements and societal needs. By analysing vast amounts of data, AI can help craft evidence-based policies that balance innovation with social equity, environmental sustainability, and public health. This capability allows policymakers to design strategies that promote scientific progress while ensuring the welfare of different communities. Moreover, AI can foster more inclusive innovation by incorporating diverse datasets and stakeholder inputs. This approach can lead to solutions that specifically address the needs of marginalised or underserved populations, ensuring that technological benefits are accessible to all. AI also plays a critical role in risk assessment and management by identifying potential unintended consequences of new technologies. Such insights can lead to the development of safeguards that protect social welfare while still advancing scientific research.

AI's ability to tailor interventions to local contexts further supports its role in bridging science and society. It can customise solutions in sectors like agriculture, healthcare, and education, making benefits more equitable and relevant. Additionally, AI-driven platforms can enhance transparency and promote public participation, allowing communities to engage actively in discussions about scientific advances and their societal implications. This openness helps build trust and align innovations with societal values.

Ethical AI development is therefore essential in ensuring that technological progress supports social welfare goals, and agricultural scientists must lead in this area. Prioritising fairness, accountability, and privacy ensures AI systems serve the greater good. While AI alone cannot resolve all tensions, it offers valuable tools to facilitate dialogue, promote responsible innovation, and develop solutions that benefit both scientific progress and societal well-being. Achieving this balance requires continuous effort, multidisciplinary collaboration, and an unwavering commitment to ethical principles.

4.2 The Contentious GM Foods

The controversy surrounding genetically modified (GM) foods is one of the most persistent and heated debates in modern agriculture, science, and politics. At the heart of this controversy are concerns over safety, environmental impact, ethical considerations, and economic influence. A significant factor amplifying these debates is the role of political lobbying by major corporations involved in biotechnology, which has led to widespread accusations of undue influence on regulatory policies and public opinion.

Proponents of GM foods argue that they offer numerous benefits, including increased crop yields, pest resistance, and tolerance to harsh environmental conditions. These technologies are seen as vital for addressing global food security challenges, especially in regions vulnerable to climate change and population growth. Supporters also contend that GM crops can reduce the need for chemical pesticides and herbicides, resulting in more sustainable farming practices. They emphasise that scientific consensus generally supports the safety of GM foods for human consumption, and that regulations in many countries include rigorous testing and assessment procedures before approval.

However, critics raise concerns over potential health risks, environmental impacts, and ethical issues. Some fear that GM foods may cause long-term health problems or allergic reactions, despite scientific evidence suggesting their safety. Environmental concerns include the potential for gene flow to wild species, the development of resistant pests, and the loss of biodiversity. Ethically, opponents question the openness of corporations in controlling seed supplies, the risk of monopolies, and the concentration of power in the hands of a few multinational corporations.

Indian institutions have developed GM brinjal, mustard, and rice. But policy caution, public opposition, trade risks, and corporate distrust have prevented their large-scale release.

A key factor fuelling this controversy is political lobbying. Large biotech companies spend significant sums on lobbying efforts to influence legislation and regulatory frameworks in their favour. Critics argue that this often results in policies that favour corporate interests over public safety and environmental health. In some cases, lobbying activities have led to the fast-tracking or loosening of regulatory standards for GM crops, bypassing thorough scientific evaluation. This has fostered suspicion and distrust among the public, who feel that decisions about food safety are driven more by corporate profit motives than by scientific evidence and public welfare.

The influence of political lobbying extends to shaping public perception and limiting the scope and transparency of scientific research. Opponents argue that disseminated information is often skewed toward positive portrayals of GM technology, while concerns and uncertainties are downplayed or ignored. This imbalance can deepen societal divisions, with some countries adopting GM crops widely, while others impose bans or strict regulations out of precaution.

While GM technology holds promise for addressing global food and environmental challenges, concerns about corporate lobbying and regulatory bias highlight the need for transparent, independent assessments and balanced policymaking. Moving forward, fostering open dialogue, promoting independent research, and ensuring that public interests are prioritised over commercial gains are essential steps to resolve this contentious debate and ensure that GM foods are developed and regulated responsibly.

4.3 The Potential of AI-powered Technologies

AI-powered technologies hold tremendous potential to revolutionise agriculture through their applications in precision farming, crop monitoring, predictive analytics, and decision support systems. These innovations enable farmers to work more efficiently, sustainably, and profitably while addressing the increasing challenges posed by climate change and resource constraints.

In the realm of precision farming and crop monitoring, AI technologies use sensors, drones, and satellite imagery to gather real-time data about soil health, moisture levels, crop growth, and pest infestations. Advanced AI algorithms analyse this vast amount of data to provide farmers with precise, location-specific insights. This targeted approach allows for optimal application of water, fertilisers, and pesticides, minimising waste and environmental impact. By continuously monitoring crop health and environmental conditions, AI-powered systems can detect early signs of diseases or pest infestations, enabling timely interventions that improve yields and reduce crop losses.

Predictive analytics and decision support systems further enhance agricultural productivity by using historical data, weather forecasts, and current field conditions to forecast future scenarios. AI models can predict yields based on various factors, helping farmers plan their harvests, manage supplies, and optimise resource use. These systems can also assess risks related to weather extremes, pests, or droughts, providing actionable recommendations for mitigating potential damages. Decision support systems synthesise this information into easy-to-understand insights, empowering farmers to make well-informed choices about planting, irrigation, fertilisation, and harvesting.

One notable example of AI-driven technology in precision farming is John Deere's See & Spray system. This innovative equipment uses computer vision and machine learning to detect weeds and pests in real time. Once identified, it precisely applies herbicides only where needed, significantly reducing chemical use and minimising environmental impact. This targeted approach not only improves crop health but also promotes more sustainable farming practices.

Another example is Climate FieldView, a digital platform that aggregates data from various sources such as soil sensors, weather stations, and satellite imagery. By analysing this data, the platform provides farmers with actionable insights and real-time field monitoring. This helps optimise planting, irrigation, and fertilisation decisions, ultimately enhancing productivity and resource efficiency. It also enables farmers to better anticipate weather-related risks and adapt their management strategies accordingly.

The Plantix app is an AI-powered mobile solution that assists farmers in quickly diagnosing crop problems. Farmers can upload images of affected plants, and the app uses advanced image recognition algorithms to identify pests, diseases, or nutrient deficiencies. It then offers tailored treatment recommendations, enabling early intervention that can save crops from severe damage. This kind of rapid, accessible diagnosis is especially valuable in regions with limited access to expert agronomists.

FarmLogs is another example of an AI-backed platform that leverages predictive analytics to forecast crop yields and track weather patterns. It also provides insights into soil health, helping farmers make better decisions about planting schedules, resource allocation, and risk management. By offering comprehensive data analysis, FarmLogs supports farmers in planning more effectively and reducing uncertainties associated with unpredictable environmental conditions.

Ecorobotix exemplifies how robotics and AI can be combined for sustainable pest and weed control. This autonomous robot uses computer vision to identify weeds and selectively sprays herbicides only where necessary. Its precision reduces chemical inputs, promotes environmental health, and decreases costs. Collectively, these AI-powered solutions demonstrate their potential to transform agriculture into a more precise, sustainable, and resilient industry.

Together, these AI-driven technologies enable smarter, more resilient farming practices. They help farmers conserve water and energy, reduce chemical usage, and adapt quickly to changing environmental conditions. By improving resource efficiency and crop health management, AI solutions contribute to sustainable agriculture that can meet global food demands while protecting the environment. As AI continues to advance, its integration into early warning systems, automated machinery, and tailored crop management strategies promises to unlock even greater productivity and sustainability in agriculture, making it more resilient and adaptive to future challenges.

4.4 Alternative innovations

In the context of modern agriculture, alternative innovations refer to new and emerging practices, technologies, and approaches that seek to address the limitations of conventional farming methods while promoting sustainability, efficiency, and resilience. These innovations aim to reduce reliance on chemical inputs, enhance resource management, and adapt to environmental challenges, often emphasising ecosystem health and social equity.

One prominent example of alternative innovation is organic farming, which avoids synthetic fertilisers and pesticides in favour of natural alternatives such as composting, crop rotation, and biological pest control. Organic practices aim to produce food that is healthier for consumers and less harmful to the environment, while also preserving biodiversity and soil vitality. Similarly, agroforestry integrates trees and shrubs into agricultural landscapes, promoting biodiversity, improving soil health, and providing additional income sources for farmers.

4.4.1 Bio-fertilisers

One notable success story of biofertilisers is their significant impact on Indian agriculture, particularly through the promotion of legacy microbial inoculants such as Rhizobium, Azospirillum, and Phosphobacteria. These biofertilisers have revolutionised the way farmers enhance soil fertility and crop productivity, especially in resource-constrained regions. In India, the use of biofertilisers has helped reduce dependence on chemical fertilisers, which are often expensive and can cause soil and water pollution over time. For example, the widespread adoption of Rhizobium inoculants in legume cultivation has led to improved nitrogen fixation, resulting in healthier crops and higher yields. One notable example is the cultivation of pulses and soybeans in states like Madhya Pradesh and Maharashtra, where the use of biofertilisers has significantly increased yields while reducing production costs. Farmers reported not only improved crop health but also long-term benefits such as increased soil organic matter and microbial diversity.

This success has contributed to sustainable agriculture practices by reducing chemical fertiliser input, lowering environmental pollution, and improving farmers' incomes. The Indian government has actively promoted the use of biofertilisers through programs such as the National Project on Organic Farming and the

Biofertiliser Development and Promotion Program, which have facilitated technology transfer, subsidy schemes, and awareness campaigns. This support has further accelerated adoption rates and demonstrated that biofertilisers can be a viable, eco-friendly alternative to chemical fertilisers, with benefits extending beyond productivity to environmental health.

4.4.2 Permaculture

Another innovative approach is permaculture, which focuses on designing sustainable and self-sufficient agricultural systems that mimic natural ecosystems. Permaculture emphasises diversity, resilience, and efficiency, encouraging farmers to develop land management strategies that require minimal external inputs and can adapt to changing conditions. In addition, regenerative agriculture aims to restore soil health through practices such as cover cropping, minimal tillage, and composting, ultimately increasing carbon sequestration, water retention, and biodiversity. Permaculture and regenerative agriculture demonstrate that the future of farming lies in working with nature, not against it—by restoring soil, nurturing biodiversity, and designing resilient systems that sustain both the land and its inhabitants.

One compelling success story of permaculture is its implementation in the Zaytuna Farm in New South Wales, Australia, led by renowned permaculture designer and educator Geoff Lawton. This project has become a global exemplar of how permaculture principles can transform degraded land into productive, self-sustaining ecosystems. At Zaytuna Farm, the goal was to rehabilitate a large area of arid, degraded land with poor soil fertility and limited water resources. Using permaculture techniques such as contour planting, water harvesting, composting, companion planting, and creating diverse polycultures, the farm gradually transformed into a lush, resilient landscape. The design focused on creating water-retentive swales, keyhole gardens, and edible food forests that stabilised the soil, improved water infiltration, and enhanced biodiversity.

Over time, the farm began producing a variety of food crops, medicinal plants, and herbs, all without the reliance on external inputs or chemical fertilisers. The project demonstrated how permaculture can restore ecological balance while providing sustainable food sources and livelihoods for local communities. Importantly, it showed that even degraded, arid environments could be turned into productive landscapes through thoughtful design and integrated systems. The success of Zaytuna Farm has inspired countless farmers, gardeners, and communities worldwide to adopt permaculture practices. It exemplifies how sustainable agriculture can be achieved in harmony with nature — conserving water, maintaining soil health, promoting biodiversity, and producing abundant food with minimal external inputs. This story highlights permaculture's potential not just as an agricultural method but as a holistic approach to creating resilient and sustainable food systems that sustain communities and restore ecosystems.

4.5 The role of AI in overcoming barriers

AI plays a crucial role in overcoming knowledge silos and barriers in agriculture by facilitating the seamless integration and sharing of information across different disciplines, institutions, and stakeholders. Traditionally, agricultural knowledge has often been fragmented, confined within specific research groups, organisations, or geographic regions, which limits innovation and the widespread adoption of best practices. AI's ability to analyse vast and diverse datasets helps break down these silos, fostering more collaborative and holistic approaches to farming.

One significant contribution of AI is its capacity to aggregate data from multiple sources, such as weather forecasts, soil sensors, satellite imagery, research studies, and farmer experiences. By synthesising this information into accessible insights, AI enables farmers, researchers, and policymakers to access a comprehensive understanding that was previously difficult to achieve. AI-powered systems facilitate knowledge sharing through tools like decision support systems, virtual advisory services, and online platforms that connect different stakeholders. These tools enable local farmers to leverage global research and technological advancements and encourage collaboration across disciplines, such as agronomy, ecology, and data science, fostering innovative solutions to complex agricultural challenges. In addition, AI enhances the capacity for predictive modelling, early warning systems, and scenario analysis, which promote proactive decision-making. By providing insights grounded in collective data, AI helps disseminate knowledge more broadly, ensuring that advancements reach even remote or underserved communities.

4.5.1 Misinformation

AI can play a vital role in eradicating misinformation by leveraging its capabilities to analyse, verify, and promote accurate information across various platforms. Through advanced algorithms and machine learning techniques, AI systems can identify false or misleading content, flag it for review, and prevent its spread, thereby improving the overall quality of information accessible to the public. One key way in which AI combats misinformation is through fact-checking tools that automatically scrutinise claims made in news articles, social media posts, and other digital content. These tools compare the claims against credible sources, databases, and verified data to assess their accuracy. When discrepancies are detected, AI can flag suspicious content or provide users with context and verified information, helping to reduce the impact of false narratives.

AI also enhances content moderation on social media and online platforms by detecting and filtering out spam, bots, and manipulated media such as deepfakes or doctored images. This approach helps safeguard against the misuse of automated accounts and AI-generated fake content, thereby reducing the rapid spread of misinformation. Moreover, AI-driven personalisation can help guide users towards reputable sources by promoting trustworthy content and reducing exposure to dubious information. For instance, recommendation algorithms can prioritise and highlight factually accurate information from verified outlets, in turn fostering informed public discourse.

However, it's important to note that AI itself must be carefully managed to prevent biases and ensure ethical use. When deployed responsibly, AI acts as a critical tool in the ongoing effort to combat misinformation, protect democratic processes, and promote an informed society. Ultimately, AI's ability to analyse vast amounts of data quickly and accurately makes it a powerful ally in eradicating false information and enhancing the reliability of digital content.

4.5.2 Scalability

AI enables scalability by allowing processes, solutions, and systems to grow efficiently to meet increasing demands without proportional increases in resources or costs. Its capacity to analyse large volumes of data rapidly and automate complex tasks makes it a powerful tool for expanding operations across different sectors, including agriculture, healthcare, manufacturing, and services. One way AI facilitates scalability is through automation. AI-powered tools can perform repetitive, time-consuming tasks such as data entry, monitoring, and decision-making with minimal human intervention. This automation allows organisations to handle a larger volume of work, serve more customers, or manage more extensive operations without requiring a proportional increase in staff or infrastructure. For example, AI-driven crop monitoring systems can oversee thousands of acres of farmland, providing real-time insights that would be impossible to manage manually at such a scale.

Additionally, AI enhances decision-making with predictive analytics and insights that support expansion. By analysing vast datasets, AI can forecast trends, optimise resource allocation, and identify new opportunities. This enables businesses and organisations to make strategic decisions confidently while scaling operations, whether that involves entering new markets, increasing production, or deploying new services. AI also enables the development of scalable platforms and solutions, such as cloud-based AI services, which can be accessed and utilised by numerous users simultaneously. These platforms remove the physical and technical barriers to growth, allowing small startups to leverage advanced AI tools without heavy investments in infrastructure. AI's capabilities in automation, data analysis, predictive modelling, and platform development make it a vital enabler of scalability. It allows organisations to expand efficiently, adapt quickly to changing demands, and innovate continuously while controlling costs and improving overall effectiveness.

4.5.3 Local Adaptation

AI can significantly facilitate local adaptation by enabling solutions that are customised to specific environmental, social, and economic conditions of different communities. Its ability to analyse diverse datasets and generate tailored insights allows farmers, policymakers, and other stakeholders to develop strategies that are more relevant and effective in their unique contexts. One key way AI supports local adaptation is through real-time data analysis and predictive modelling. For example, AI-powered climate models can forecast weather patterns at a granular level, enabling farmers to adjust their planting schedules, select suitable crop varieties, and implement water management practices tailored to their local climate conditions. Similarly, AI-driven soil and crop monitoring tools can provide specific recommendations based on local soil health, crop needs, and pest pressures, optimising resource use and reducing risks of crop failure.

AI also facilitates the development of localised decision support systems that incorporate traditional knowledge, local practices, and community-specific data. These systems help farmers make informed choices about irrigation, fertilisation, pest control, and harvesting, aligning interventions with local realities. Moreover, AI-powered platforms can analyse socioeconomic data to support policy interventions that address local challenges, such as food security, land rights, and resource management. Another aspect of AI's role in local adaptation is in developing and promoting culturally appropriate technologies. By involving local communities in the design and deployment of AI solutions, technology can be tailored to meet their language, literacy levels, and social norms, increasing acceptance and effectiveness.

AI is thus breaking down knowledge silos in agriculture by integrating data from weather, soil, satellites, research, and farmer experiences into actionable insights. It enables smarter, location-specific decisions, fosters collaboration across disciplines, and powers tools like advisory platforms and predictive models. By connecting local farmers with global innovations, AI enhances agriculture's resilience, sustainability, and ability to address food security challenges.

Summary

This chapter invites us to rethink traditional technological fixes in agriculture—whether GMOs, biofertilisers, or even AI—by embracing the idea of Alternative Innovation. It calls for a shift away from uniform, one-size-fits-all solutions toward living, localised approaches that honour the unique soil, water, and cultural rhythms of each farming community. In this vision, AI becomes not just a tool, but a partner to nature and farmers, capable of sensing the needs of plants and the pulse of the environment in real time. By guiding precise and timely actions, AI can help create resilient, regenerative farming systems that sustain both the land and the people who depend on it. This is not just about smarter agriculture—it is about cultivating harmony between technology, tradition, and the Earth itself. True innovation in agriculture is not about imposing uniform solutions, but about weaving technology into the fabric of local wisdom—so that every field, every farmer, and every ecosystem can thrive in harmony.

Academic Exercise

1. What is meant by the gap between science and social welfare?
2. Give examples of the potential of AI-powered technologies in agriculture.
3. Explain the term "alternative innovation" with examples.
4. Define Permaculture.
5. Describe AI as an enabler of smallholder.

5

Extension 2.0

> ***Today's extension worker is no longer just a messenger of science, but a leader of change—bridging policy, people, and perception to shape the future of farming.***

5.1 Persuasion as Communication

Agriculture extension is a dynamic, resource-based system that involves providing farmers and rural communities with technical knowledge, practical skills, and new technologies to improve agricultural productivity, sustainability, and income. It serves as a bridge between research institutions and farmers, facilitating the dissemination of innovative farming practices, weather forecasting, crop management, pest control, and market information. The goal of agricultural extension is to equip farmers with the information and skills necessary to improve their productivity and livelihoods, while promoting sustainable agricultural practices.

Historically, the extension tradition began in the early 20th century. It is rooted in the idea of providing farmers with knowledge, advice, and support to enhance their farming methods and improve their livelihoods. This tradition generally emphasises personal contact, farmer education, and the dissemination of research-based information through various approaches such as field demonstrations, workshops, and farmer training programs influenced by government initiatives aimed at modernising agriculture, especially undertaken during the Green Revolution. It has since evolved to incorporate new technologies, communication methods, and participatory approaches, focusing on farmer needs, sustainability, and community development.

In the context of extension work or marketing, this evolution signifies a shift from merely delivering facts to engaging with the audience in a way that motivates them to act. Effective persuasion involves understanding the audience's needs, using credible sources, emotional appeal, and logical reasoning to foster change or acceptance. This approach recognises that influence plays a crucial role in shaping opinions, encouraging adoption of new practices, or changing resistant behaviours, making communication a powerful tool for social, behavioural, or attitudinal change. Influence shaping opinions involves guiding and moulding individuals' attitudes, beliefs, and perceptions through strategic communication and persuasive techniques. This process begins with a deep understanding of the target audience's values and concerns, so that messages can be crafted in a manner that resonates effectively.

Credible sources and trusted messengers enhance the persuasion, making the message more convincing. Emotional appeals, such as invoking hopes or fears, can motivate individuals to reconsider or reinforce their viewpoints, while logical evidence provides rational support to sway opinions through facts and reasoning. Repetition and consistency strengthen influence over time, embedding the desired perspectives more firmly.

Additionally, demonstrating that peers or respected figures support a particular view—known as social proof—can significantly impact opinions by encouraging conformity. Overall, influence shaping opinions combines emotional, logical, and social elements to alter attitudes and perceptions, ultimately guiding actions and decisions in a direction aligned with specific goals.

In modern contexts, especially in fields like agriculture, marketing, and public policy, effective communication involves the skilful use of emotional appeal, credible messaging, and strategic storytelling to motivate action and foster trust. This shift emphasises the importance of understanding the audience's needs and perspectives, using evidence-based arguments combined with persuasive techniques to create resonance and drive change. Ultimately, persuasion as an art of communication aims to inspire confidence, shape perceptions, and encourage positive decisions, making it an essential skill for leaders, educators, and policymakers in today's interconnected world.

5.2 Leadership Role of Influencing Policy and Public Perception

In agriculture, leadership plays a crucial role in influencing policy and shaping public perception to promote the sector's growth and sustainability. Agricultural leaders, including policymakers, extension officers, and industry stakeholders, advocate for supportive policies that address farmers' needs, improve agricultural practices, and ensure fair resource distribution. They work to influence government decisions by providing expert insights, data, and advocacy efforts aimed at creating conducive environments for agricultural development. Simultaneously, agricultural leaders shape public perception by educating communities about the importance of agriculture, sustainable practices, and food security. Through effective communication and advocacy, they foster greater awareness, dispel misconceptions, and garner public support for agricultural initiatives. This dual influence on policy and perception is vital for mobilising resources, encouraging innovation, and ensuring that agriculture receives the attention and investment necessary for long-term prosperity.

In India, agricultural policies are primarily formulated and made by the central government and state governments. The main body responsible at the national level is the Ministry of Agriculture and Farmers' Welfare. This ministry develops broad policy frameworks, strategies, and programs related to agricultural development, food security, and rural welfare. At the state level, the respective State Departments of Agriculture work within the guidelines set by the central government but tailor policies to local needs and conditions. The process of policy-making often involves consultation with various stakeholders, including farmers' organisations, agricultural experts, research institutions, and industry representatives. Additionally, policy decisions are influenced by parliamentary debates, committees, and advisory bodies, ensuring a collaborative approach to shaping India's agricultural landscape.

The role of extension officers is vital in bridging the gap between research institutions, government policies, and farmers' practical needs. They serve as the primary link in delivering agricultural knowledge, technologies, and support services to rural farmers. Extension officers assess local farming conditions, identify farmers' needs, and then disseminate relevant information on modern farming techniques, crop management, pest control, soil health, and weather forecasts. They also organise training programs, demonstration plots, and field days to illustrate new practices and innovations. Moreover, extension officers provide personalised advice and troubleshooting assistance, helping farmers adopt sustainable and profitable farming methods. They play a key role in promoting agricultural reforms, policies, and government schemes, ensuring farmers are aware of and can access these resources. Additionally, extension officers act as a feedback channel, relaying farmers' concerns and challenges to policymakers and researchers for continuous improvement of agricultural practices and policies. Their work enhances productivity, encourages the adoption of new technologies, and contributes to rural development and food security.

As Dr M. S. Swaminathan observed, 'Agriculture is the backbone of the livelihood security system of nearly 700 million people,' highlighting why bridging research and farming through effective extension is more than a support function—it is essential to national food and livelihood security.

Industry stakeholders influence farmers in several significant ways, shaping their decisions and practices to align with market demands and technological advancements. These stakeholders include input suppliers (such as seed, fertiliser, and pesticide companies), agro-processing industries, traders, exporters, financial institutions, and technology providers. They influence farmers through the provision of high-quality inputs, access to new technologies, and innovative farming practices that can increase productivity and profitability. Industry stakeholders also influence farmers by creating market opportunities and establishing supply chains that motivate farmers to adopt specific crops or agricultural practices to meet market standards and consumer preferences. Additionally, they often provide credit facilities, insurance, and extension support, further guiding farming decisions. Marketing campaigns, contractual agreements, and capacity-building programs conducted by these stakeholders also play a role in shaping farmers' perceptions and choices, ultimately impacting productivity, cropping patterns, and income levels. Overall, industry stakeholders serve as key drivers in the modernisation and commercialisation of agriculture.

5.3 Effective communication, engagement, and advocacy

In the context of modern agriculture, effective communication, engagement, and advocacy are essential for promoting innovation, sustainable practices, and policy support. Effective communication involves clearly conveying complex technical information, new technologies, and policy initiatives in a manner that farmers, stakeholders, and the public can understand and act upon. Engagement entails actively involving farmers and local communities in decision-making processes, fostering participatory approaches that consider their needs, preferences, and feedback. This two-way interaction builds trust, enhances adoption of new practices, and strengthens collaboration among all stakeholders.

The multiplicity of languages in India significantly impacts effective communication, engagement, and advocacy in agriculture. India is a linguistically diverse country, with over 1,600 languages spoken across different regions. This linguistic diversity poses challenges in disseminating agricultural information uniformly and effectively. When extension services, training programs, and policy messages are delivered only in a few dominant languages, many farmers—especially those in remote or rural areas—may not fully understand or be able to access the information, limiting the adoption of new practices or technologies.

To overcome this, agricultural communication strategies need to be culturally and linguistically sensitive, using regional languages, local dialects, and visual communication methods such as demonstrations and illustrations. Employing local influencers, community leaders, and multilingual extension workers can foster better engagement and trust. Moreover, addressing language barriers enhances inclusivity, ensuring that all farmers, regardless of linguistic background, have equal access to vital information. In summary, the multiplicity of languages requires tailored communication approaches to ensure that agricultural extension, advocacy, and engagement efforts reach and effectively impact diverse farmer communities across India.

Advocacy plays a crucial role in influencing policies, attracting investments, and raising awareness about the importance of sustainable agricultural practices, climate resilience, and food security. By championing these issues, agricultural leaders and extension workers can mobilise support at local, national, and international levels. Together, these roles help create an informed and motivated agricultural community capable of embracing change, driving innovation, and ensuring the sector's long-term sustainability and prosperity.

An example of advocacy in agriculture is a campaign by environmental organisations and farmers' groups to promote the adoption of organic farming practices. This advocacy effort aims to influence government policies by raising awareness about the benefits of organic agriculture for soil health, crop productivity, and environmental sustainability. The groups organise public seminars, submit policy briefs to lawmakers, and conduct media campaigns highlighting successful organic farming models. They also advocate for government subsidies, training programs, and certification support to encourage more farmers to transition to organic methods. Through these activities, the advocacy aims to create favourable policies, shift public perception positively towards organic farming, and ultimately support sustainable agricultural development.

5.4 Addressing misinformation with data-driven insights

In the digital age, access to vast amounts of information has revolutionised how we learn, communicate, and make decisions. However, this abundance of information has also led to the proliferation of misinformation—false or misleading information that can spread rapidly through social media, news outlets, and other digital platforms. Misinformation poses significant challenges across various sectors, including public health, politics, and agriculture, where inaccurate data can lead to poor decision-making, fear, and societal discord. To effectively combat misinformation, leveraging data-driven insights has emerged as a crucial strategy, providing evidence-based approaches to identify, analyse, and counter false narratives.

Data-driven insights involve the collection, analysis, and interpretation of large datasets to uncover patterns, trends, and correlations. In the context of misinformation, this approach enables organisations and policymakers to detect the early signs of false information spreading online. For example, social media analytics can identify viral false claims by monitoring keywords, hashtags, and engagement metrics. By employing machine learning algorithms and natural language processing techniques, it is possible to classify content as credible or suspicious, allowing for rapid response and targeted intervention.

Furthermore, data-driven insights help understand the source, reach, and impact of misinformation. This understanding is critical for designing effective countermeasures. For instance, if data analysis reveals that a particular false claim is gaining widespread attention among specific demographics or regions, targeted communication campaigns can be developed to address those audiences directly. These campaigns can include credible information, expert opinions, and factual evidence tailored to dispel myths and misconceptions.

In addition to detection, data-driven insights support the development of reliable fact-checking systems. Automated fact-checkers equipped with databases of verified information can cross-reference claims in real-time, highlighting inaccuracies and providing correct data. This not only helps in correcting the misinformation but also builds public trust in factual content. Moreover, transparency in data sources and methodologies enhances credibility, encouraging the public to rely on scientifically verified information.

In the agricultural sector, for example, misinformation about modern farming techniques or the safety of genetically modified crops can hinder adoption and sustainability efforts. By analysing farmer feedback, social media discussions, and research data, agricultural extension services can proactively address misconceptions with factual insights. Providing farmers with evidence-based guidelines and success stories can promote informed decision-making and reduce the spread of misinformation. As the digital landscape continues to evolve, embracing data-driven strategies will be essential for safeguarding truth and promoting informed decision-making across all sectors.

5.5 Counteracting lobbying through digital diplomacy

In an increasingly interconnected world, lobbying has become a common tool used by various organisations, interest groups, and nations to influence public policy and decision-making processes. While lobbying can serve legitimate purposes, such as advocating for consumer rights or environmental protection, it can also be misused to sway policies in favour of special interests, sometimes at the expense of public welfare or democratic principles. Counteracting undue lobbying influence has become a critical challenge, and digital diplomacy offers an effective way to address this issue by promoting transparency, fostering informed public discourse, and engaging stakeholders directly through digital platforms.

Digital diplomacy refers to the use of digital tools and social media technologies by governments, institutions, and civil society to communicate, negotiate, and promote their interests in an open and transparent manner. One of its key roles in counteracting lobbying is enhancing transparency and accountability. Governments and organisations can use digital platforms to share detailed information about policy-making processes, disclose lobbying activities, and explain the rationale behind decisions. By making this information accessible to the public, digital diplomacy prevents opaque lobbying practices from dominating policy debates, ensuring that citizens are better informed and able to participate meaningfully.

Moreover, digital diplomacy enables direct engagement with civil society and the general public, creating spaces for open dialogue and debate. Governments and policymakers can use social media, online forums, and webinars to gather feedback, address concerns, and dispel misinformation propagated by vested interests. This interactive approach ensures that diverse perspectives are considered, and the influence of powerful lobbying groups is balanced by grassroots voices and expert opinions. For example, international organisations and governments can run awareness campaigns highlighting the importance of equitable policymaking and exposing undue influence by special interests.

Another vital aspect of digital diplomacy is to counter-misinformation and promote credible, evidence-based information. Interest groups often deploy sophisticated campaigns to sway public opinion and distort facts. Digital diplomacy can serve as a countermeasure by disseminating accurate data, expert analysis, and transparent information through official channels. Social media campaigns, fact-checking initiatives, and online educational programs strengthen public trust in authentic information sources and reduce the impact of manipulative lobbying efforts.

Furthermore, digital diplomacy fosters robust international cooperation and knowledge-sharing, which is crucial in addressing cross-border lobbying practices that may undermine global efforts on issues such as environmental protection, human rights, and public health. By collaborating across nations through digital platforms, governments can develop coordinated strategies to resist unfair lobbying influence and promote policies rooted in collective interest and sustainable development. As technology continues to evolve, digital diplomacy will play an increasingly pivotal role in shaping equitable and accountable governance, ultimately strengthening democracy in the digital era.

5.6 New Age Agriculture Extension Worker

A 21st-century agriculture extension worker is a dynamic, tech-savvy professional dedicated to empowering farmers through innovative, inclusive, and sustainable practices. They possess a strong understanding of modern agricultural technologies, environmental sustainability, and climate resilience, enabling them to provide up-to-date information and practical solutions. These extension workers often utilise digital tools such as mobile apps, social media, and online platforms to reach a broader audience, especially in remote and underserved areas. They are skilled communicators, capable of translating complex scientific research into simple, actionable advice tailored to farmers' needs. Beyond technical knowledge, they exhibit qualities of adaptability, cultural sensitivity, and problem-solving, working closely with diverse stakeholders, including farmers, government agencies, research institutions, and private sector players. They also prioritise gender inclusivity, youth engagement, and participatory approaches to ensure equitable access to resources and knowledge. A 21st-century agriculture extension worker is indeed a catalyst for agricultural modernisation, sustainability, and rural development, committed to fostering resilient and productive farming communities through innovative practices and continuous learning.

This involves supporting farmers to develop the ability to withstand, adapt to, and recover from challenges such as climate change, pests, economic fluctuations, and resource scarcity. It involves promoting sustainable farming practices, diversifying income sources, and implementing innovative technologies that enhance productivity while ensuring environmental health. Building resilience also includes providing farmers with access to knowledge, financial resources, and social support systems that help them navigate risks and uncertainties. The goal is to create farming communities that are not only capable of maintaining or increasing their agricultural output but are also adaptable to changing conditions, thereby ensuring their long-term sustainability, food security, and overall well-being.

Local actions, for example, can play a crucial role in addressing climate change by implementing targeted strategies that reduce greenhouse gas emissions, improve resilience, and promote sustainable development within communities. Some effective local actions include promoting renewable energy sources such as solar and wind power to replace fossil fuels, encouraging energy-efficient practices in households and businesses, and enhancing waste management through recycling and composting to reduce methane emissions. Additionally, protecting and restoring local natural ecosystems, such as forests, wetlands, and green spaces, can absorb carbon dioxide and improve climate resilience.

Local governments can also promote sustainable transportation options like cycling, walking, and public transport to reduce emissions from vehicles. Adapting infrastructure to withstand extreme weather events—such as flood defences and resilient building designs—helps communities recover faster from climate impacts. Educating residents about climate change and encouraging responsible consumption and conservation practices fosters a culture of sustainability. Engaging local stakeholders—such as farmers, businesses, youth groups, and civil society—ensures that climate action planning solutions are practical, inclusive, and tailored to the community. By taking these collective actions, local communities can make a significant contribution to global efforts to mitigate climate change while safeguarding their own environment, economy, and public health.

5.7 Elevating Extension into a Leadership Role

Raising the profile of extension into a leadership role entails transforming conventional agricultural extension practitioners into strategic leaders who spearhead innovation, shape policies, and motivate sustainable developments within their communities. This transition requires extension workers to develop a broad skill set that includes not only technical expertise but also leadership qualities such as vision, strategic thinking, decision-making, and effective communication.

When agricultural extensionists rise as leaders, they cease to be mere messengers of technology and become architects of change—visionaries who inspire innovation, shape policies, and guide communities toward sustainable futures.

In a leadership role, extension professionals serve as change agents, facilitating the adoption of new technologies, fostering participatory development, and guiding communities through complex challenges such as climate change, resource management, and market access. They also collaborate with policymakers, research institutions, NGOs, and the private sector to create an environment conducive to agricultural growth and rural development.

Furthermore, elevating extension into leadership involves championing inclusive practices that empower marginalised groups such as women and youth, promoting sustainable and climate-resilient agriculture, and advocating for policies that benefit small-scale farmers. It also means being proactive in shaping regional and national agricultural policies, mobilising resources, and inspiring innovation at the grassroots level. By positioning extension as a leadership force, agricultural sectors can accelerate progress towards food security, sustainability, and rural prosperity, ensuring that extension professionals serve not just as advisors but as influential leaders and catalysts for positive change.

Dr M.S. Swaminathan played a pivotal role in transforming Indian agriculture through scientific research, policy advocacy, and leadership in agricultural extension. His efforts helped increase food production and promote sustainable farming practices. Dr A.P.J. Abdul Kalam, although primarily a scientist and former President of India, contributed significantly to rural and agricultural development through his visionary leadership in promoting science and technology-based farming solutions.

Kisan Swabhiman, the Farmers' Self-Respect Movement, has played a leadership role in advocating for farmers' rights, influencing agricultural policies, and promoting sustainable farming practices. Krishi Vigyan Kendras (KVKs) are leading institutions in India that serve as hubs of extension leadership at district levels. KVK heads and scientists act as extension leaders by providing farmer education, demonstrating new technologies, and influencing local agricultural practices. These examples reflect diverse forms of extension leadership in India, ranging from individual visionaries to institutional and community-based leaders, all contributing to the development of sustainable and progressive agriculture.

The transformation of Indian agriculture will not be achieved by policy alone, but by people — especially those who stand closest to the farmers. When agricultural extensionists evolve from intermediaries into leaders, they become the bridge between knowledge and empowerment, policy and practice. Their leadership can dismantle exploitative systems, rekindle trust, and restore balance to the rural economy. The next revolution in agriculture, therefore, will not come from laboratories or legislatures, but from the hearts and minds of extension leaders who walk with farmers, transforming struggle into strength and subsistence into self-reliance.

Summary

This chapter explores the evolution of agricultural extension into a modern, leadership-oriented discipline. It highlights how communication has transformed into an art of persuasion, emphasising the importance of shaping public perception in today's media-savvy world. The chapter underscores the crucial role of advocacy in translating policy into ground-level action, ensuring that policies effectively benefit farmers and communities. It also discusses the importance of countering misinformation through data-driven evidence, strengthening credibility and building trust among stakeholders. Digital diplomacy is presented as a strategic tool for fostering transparency, engagement, and international cooperation in agriculture. The profile of the contemporary extension worker is outlined, emphasising their role as a proactive leader capable of driving change. Overall, the chapter advocates for viewing extension professionals as leaders and change-makers, essential for shaping sustainable, informed, and resilient agricultural communities in the 21st century.

Academic Exercise

1. How does communication evolve into persuasion?
2. Describe the Indian agriculture policy-making chain.
3. Explain the term "alternative innovation" with examples.
4. Profile a new-age agriculture extension worker.
5. How is extension elevated into a leadership role?

6

Values-Driven Leadership

In the age of AI, leadership's true compass is not efficiency, but ethics—guiding technology to serve humanity and the earth.

6.1 Navigating with a Moral Compass

A moral compass is a person's internal sense of right and wrong that guides their behaviour and decision-making. It helps individuals navigate ethical dilemmas and determine what they believe to be just, fair, and morally acceptable. Values-Driven Leadership is a comprehensive leadership philosophy that emphasises guiding actions, decisions, and behaviours based on a set of core values and ethical principles. A value system acts as a moral compass. It consists of the principles and beliefs that a person or group considers important, guiding their judgments and actions in situations involving ethical considerations.

Leaders who adopt this approach view their moral beliefs—such as integrity, honesty, respect, fairness, responsibility, and compassion—as fundamental to their role. These values serve as a moral compass, helping leaders navigate complex situations with consistency and purpose. By anchoring their leadership style in these principles, they aim to foster an environment of trust, credibility, and ethical integrity. A system replaces the whims and fancies of those in power, aiding them in navigating complex moral decisions and maintaining alignment with their core values.

At the heart of Values-Driven Leadership is ethical decision-making. Leaders evaluate their choices against their core values, ensuring that their actions align with the organisation's mission and moral standards. This commitment to authenticity and integrity not only builds credibility but also sets a powerful example for others within the organisation. When leaders consistently act in ways that reflect their stated values, they create a culture of trust, accountability, and mutual respect. This foundation encourages employees to adopt similar principles, leading to a more cohesive and morally aligned organisational environment.

Furthermore, Values-Driven Leadership is instrumental in shaping a positive organisational culture. Leaders who lead by example foster an atmosphere where collaboration, innovation, and high ethical standards flourish. Over time, this approach cultivates a strong sense of shared identity and purpose among team members. It also emphasises the importance of long-term success rooted in moral principles, rather than short-term gains. By maintaining a focus on social responsibility and sustainability, organisational trust and reputation are enhanced, which can contribute to resilience during difficult times.

Implementing Values-Driven Leadership involves ongoing commitment and clear communication of core principles. Leaders must embed these values into every aspect of organisational life—policy formulation, everyday interactions, and strategic planning.

Challenges may arise in balancing competing values or external pressures, but a steadfast commitment to moral clarity helps to navigate these dilemmas. When effectively practised, this leadership approach empowers employees at all levels to embrace and uphold organisational values, creating an environment where ethical behaviour is the norm. In summary, Values-Driven Leadership leads organisations on a path of moral purpose, fostering trust, integrity, and sustainable success.

One of the most prominent examples of values-driven leadership is Mahatma Gandhi, whose entire philosophy and practice of leadership were anchored in timeless principles—non-violence (Ahimsa), truth (Satya), and justice. For Gandhi, these were not abstract ideals but living disciplines that shaped every decision he made and every strategy he employed in India's struggle for independence. His steadfast refusal to compromise on moral grounds gave him a rare authenticity that resonated deeply with people, earning their trust and commanding respect across divides of class, caste, and creed. Gandhi transformed resistance into a moral movement, showing that courage rooted in truth could be stronger than armies and that dignity could prevail even without weapons. By embodying integrity in both personal conduct and public action, he mobilised millions, inspired global movements for civil rights and freedom, and proved that leadership grounded in ethical clarity can achieve change that endures far beyond one lifetime.

Chaudhary Charan Singh exemplified values-driven leadership through his commitment to rural development, social justice, and the rights of farmers. As a champion of the agrarian community, he dedicated his career to advocating for the needs of farmers and rural populations, emphasising values such as justice, fairness, and compassion. He worked tirelessly to implement policies that aimed to uplift farmers, reduce exploitation, and promote land reforms. His focus on these core values guided his actions, even in the face of political opposition and challenges, demonstrating his integrity and commitment to his principles.

A nation can prosper only when its rural sector is upgraded and its purchasing power is high.

Chaudhary Charan Singh

Prime Minister Narendra Modi emphasises core values like nationalism, development, and social service in his leadership. His focus on initiatives such as Digital India, Swachh Bharat (Clean India), and Make in India demonstrates a drive to align policies with patriotic and development-oriented values. While his leadership has generated both praise and controversy, his emphasis on national pride and social upliftment exemplifies a values-driven approach centred around India's collective progress.

American agronomist Norman Borlaug, botanist George Washington Carver, ecologist William J. Beal, horticulturist Luther Burbank, and our own M.S. Swaminathan were all great value-driven leaders, in addition to their scientific excellence. Sir John B. H. S. Briscoe was an influential agricultural scientist in Australia. Dr Verghese Kurien and Dr R.S. Paroda prepared a generation of young Indian scientists to do socially relevant scientific work.

6.2 Balancing Innovation and Responsibility

Balancing innovation and responsibility is indeed the need of the hour in today's fast-paced and interconnected world. Innovation drives progress, economic growth, and competitive advantage, enabling organisations and societies to solve complex problems, improve the quality of life, and stay ahead in a rapidly changing landscape. However, without a sense of responsibility, innovation can lead to unintended consequences such as ethical breaches, environmental harm, and social inequality.

The challenge lies in fostering innovative ideas while ensuring they are aligned with ethical standards and social values. Organisations must develop a culture where creativity and risk-taking are encouraged, but within frameworks that prioritise sustainability, transparency, and social good. For instance, technology companies developing artificial intelligence or biotech innovations are increasingly bearing the responsibility to create safe, ethical, and inclusive solutions that benefit society rather than harm it. In the Indian context, this balance is critical as the country navigates rapid technological development, economic reforms, and social change.

Policies promoting green energy, responsible use of technology, and inclusive growth exemplify efforts to integrate responsibility with innovation. Leaders in government, business, and civil society need to work collaboratively to ensure that innovation drives progress without compromising ethical standards, environmental sustainability, or social equity. Ultimately, the most impactful and sustainable innovations are those that are driven by a sense of responsibility towards society and the environment. Striking this balance is crucial for ensuring that technological and economic advancements serve humanity as a whole, fostering long-term prosperity and moral integrity. In business parlance, it is also called the triple bottom line – profit, people, planet - approach. Business organisations can successfully integrate economic gains with social equity and environmental sustainability, embodying the principles of the triple bottom line approach to foster long-term, holistic success.

6.2.1 Tata Group

Tata exemplifies the triple bottom line by integrating profitability with social responsibility and environmental sustainability. For instance, Tata Steel's operations focus not only on financial performance but also on community development and environmental stewardship. Their initiatives include reducing carbon emissions, investing in renewable energy, and supporting local communities through education and health programs, demonstrating how a corporation can pursue profit while caring for society and the environment.

6.2.2 ITC Limited

ITC has adopted a holistic approach to the triple bottom line by embedding sustainability into its core business model. Through its Responsible Business practices, the company has invested heavily in agricultural sustainability, empowering farmers with initiatives such as e-Choupal and promoting climate-resilient farming practices. Its large-scale afforestation and watershed development projects demonstrate a deep commitment to environmental stewardship, while innovations in eco-friendly packaging reflect a push toward circular economy solutions. At the same time, ITC has fostered rural livelihoods by supporting entrepreneurship, women's empowerment, and community development programs. Balancing these efforts with robust financial performance, ITC has demonstrated that economic growth, social equity, and environmental protection can reinforce one another, setting a benchmark for responsible corporate leadership in India.

6.2.3 Patanjali Ayurveda

When profit overshadows farmers' welfare, conflicts of interest corrode the foundations of agriculture; only transparency, fairness, and the voices of farmers can restore balance.

Patanjali, founded by Baba Ramdev, has created a significant impact by aligning its business growth with wellness, sustainability, and rural empowerment. By reviving Ayurveda and promoting organic, chemical-free products, the company has reshaped consumer preferences toward healthier lifestyles while reducing ecological footprints. Its sourcing strategy supports thousands of local farmers, ensuring fair returns and encouraging sustainable agricultural practices. At the same time, Patanjali's ability to deliver affordable, natural alternatives has disrupted traditional FMCG markets, demonstrating that profitability can coexist with social good and environmental stewardship. Patanjali's model demonstrates that true business success is measured not only in profits but in the health of people, the resilience of the planet, and the prosperity of communities.

6.3 The role of integrity, transparency, and social responsibility

Integrity refers to the quality of being honest, transparent, and consistent in moral principles and actions. It involves adhering to ethical standards and showing trustworthiness in all interactions. In values-driven leadership, integrity plays a crucial role by fostering trust, credibility, and respect among team members and stakeholders. Leaders with integrity set a moral example, align their actions with their core values, and cultivate an organisational culture of honesty and accountability. This commitment to integrity enhances ethical decision-making, promotes a positive work environment, and sustains long-term success based on trust and moral consistency. For example, an agribusiness company tests its crops for pesticides and honestly reports the findings to regulators and consumers, even if the results reveal issues that could harm their reputation, demonstrating honesty and ethical standards.

Transparency is the act of openly sharing information, intentions, and decision-making processes with others. It involves honesty, clear communication, and accountability, allowing stakeholders to understand how and why decisions are made. In values-driven leadership, transparency plays a vital role by building trust and credibility. It encourages openness, fosters dialogue, and ensures that leaders are accountable for their actions. The company provides clear and accessible information about its supply chain practices, including how it sources raw materials, pesticide use, and labour conditions, allowing consumers to make informed choices.

Social responsibility is the obligation of organisations and individuals to act ethically and contribute positively to society and the environment. It involves making decisions that benefit not only the organisation but also the broader community, including actions that promote sustainability, fairness, and social well-being. In values-driven leadership, social responsibility is essential because it reflects a commitment to ethical principles and the greater good. Leaders who prioritise social responsibility integrate these values into their strategies and decision-making, demonstrating care for stakeholders and the environment. The agribusiness must invest in sustainable farming practices that protect soil health and water resources, support local farmers through training programs, and ensure fair wages and safe working conditions for employees, contributing positively to the community and environment.

An agriculture leader guided by integrity, transparency, and social responsibility cultivates not only healthy crops but also a sustainable future and trust within the community. By integrating these core values, such a leader not only drives innovation and productivity but also fosters a culture of trust, sustainability, and ethical responsibility. This holistic approach ensures that agricultural progress benefits current and future generations, safeguarding natural resources, supporting local communities, and maintaining ethical standards that uphold the dignity and well-being of everyone involved in the industry.

6.4 Protecting smallholders and marginalised communities

A leader driven by protecting smallholders and marginalised communities is committed to advocating for their rights, well-being, and sustainable development. Such a leader actively promotes fair access to resources, fair pricing, and inclusive policies that empower these vulnerable groups. They prioritise ethical practices, social equity, and community engagement to reduce disparities and enhance their economic and social resilience. Such leaders live and work with their followers and therefore never get disconnected.

Fair access to resources refers to the equitable and just distribution of essential assets such as land, credit, technology, training, and markets, ensuring that smallholders and marginalised communities have the same opportunities to improve their livelihoods. It involves removing barriers that prevent these groups from benefiting equally from resources necessary for sustainable farming and economic growth. For example, an agricultural cooperative program provides smallholder farmers with affordable access to modern farming equipment, quality seeds, and credit facilities. The government or a development organisation ensures these resources are distributed fairly, avoiding favouritism or monopolisation by large corporations. As a result, smallholders can increase their productivity, improve income stability, and compete more effectively in the market, leading to inclusive growth and poverty reduction in marginalised communities.

Fair pricing refers to setting prices for agricultural products that ensure smallholders and marginalised communities receive just compensation for their produce, covering their costs and providing a reasonable profit. It aims to prevent exploitation by middlemen, large corporations, or market fluctuations that often disadvantage small-scale farmers. A fair-trade certification organisation establishes a minimum price for cocoa farmers' harvests, ensuring they receive a living wage regardless of market volatility. This fair price helps smallholders cover their costs, invest in sustainable farming practices, and support their families. By guaranteeing fair compensation, the initiative promotes economic stability, encourages ethical trading practices, and empowers smallholders to improve their livelihoods while aligning with social responsibility and equity.

The National Rural Livelihood Mission (NRLM), also known as Ajeevika, is a shining example of inclusive policies that empower smallholders and marginalised communities. NRLM aims to alleviate rural poverty by promoting self-employment and organisation among marginalised groups, including small-scale farmers, tribal communities, and women. The program provides training, access to micro-credit, and support for sustainable livelihood activities such as organic farming, handicrafts, and small enterprises. It encourages the formation of SHGs (Self-Help Groups), which help these communities access resources, negotiate better market prices, and gain financial independence. The policy also emphasises gender equality and participation of marginalised groups in decision-making, fostering social inclusion and economic empowerment.

Protecting smallholders and marginalised communities is the most powerful foundation for strengthening a nation, for they are the roots that sustain sustainable growth, resilience, and shared prosperity for all. When their rights are protected and their livelihoods supported, it leads to increased food security, poverty reduction, and social stability. Strengthening these groups creates a more inclusive economy where benefits are broadly shared, fostering resilience against economic shocks and environmental challenges. As Mahatma Gandhi once reminded us: "The soul of India lives in its villages." By honouring and empowering smallholders, we not only safeguard the nation's food systems but also uphold the dignity and wisdom of those who nurture the land.

6.5 Navigating Conflicts and Bias

Agriculture forms the foundation of India's economy, providing livelihoods for millions of farmers and playing a crucial role in ensuring food security across the country. Digital technologies have revolutionised the sector through data-driven solutions, including precision farming, AI-powered forecasting, and digital marketplaces. However, these innovations also bring challenges related to conflicts of interest, data privacy, and algorithmic bias, which need careful navigation to ensure inclusive, ethical, and sustainable development.

Conflicts of interest arise when stakeholders in the agricultural value chain prioritise personal or corporate gains over farmers' welfare. For example, agritech companies might promote certain seeds or chemical inputs that benefit their sales but may not be in the best interest of smallholders or the environment. Additionally, government policies aimed at promoting specific crop types or input providers might favour corporations at the expense of sustainable practices or smallholder needs. Navigating these conflicts requires robust policy frameworks, transparency, and the involvement of farmers in decision-making processes, ensuring that the benefits of technological advancements do not sideline the interests of vulnerable farmers.

When profit overshadows farmers' welfare, conflicts of interest corrode the foundations of agriculture; only transparency, fairness, and the voices of farmers can restore balance.

The increasing use of digital platforms collects vast amounts of data on farmers' behaviours, landholdings, crop patterns, and financial dealings. While this data can enhance service delivery—such as personalised advice or weather forecasts—it also raises significant privacy concerns. Farmers, especially from marginalised communities, may lack awareness or control over their data, risking exploitation or misuse. For instance, private companies may access farmers' land or financial information without their explicit consent, leading to discriminatory practices or a loss of privacy. Addressing these issues calls for the development of robust data protection laws, informed consent protocols, and transparent data-sharing policies that prioritise farmers' rights.

Algorithms used in agriculture—such as predictive models, credit scoring, or resource allocation systems—may inadvertently incorporate biases based on skewed data or flawed assumptions. This can result in discrimination against marginalised groups or smallholder farmers. For example, an AI system might overlook small farmers with limited digital literacy or inadequate data points, unjustly denying them credit or access to resources. In India, where socio-economic disparities are pronounced, algorithmic bias can reinforce existing inequalities unless actively addressed. Ensuring fairness requires diverse and representative data, continuous monitoring of AI outputs, and inclusive stakeholder participation in the design of these systems.

India has taken significant steps to help poor farmers and labourers receive financial assistance directly, avoiding middlemen who often exploit them. One notable initiative is the Pradhan Mantri Jan Dhan Yojana (PMJDY), launched in 2014, which aimed to provide universal access to banking services. Through this scheme, millions of farmers and labourers opened bank accounts in their own names, enabling direct access to financial resources. Complementing this is the Direct Benefit Transfer (DBT) system, which delivers government subsidies and benefits—such as agricultural support, pensions, and social assistance—straight into beneficiaries' bank accounts using biometric identification and Aadhaar linkage.

Programs like PM-KISAN represent a significant step toward strengthening farmers' economic security by ensuring that financial assistance is transferred directly into their bank accounts, without leakage or diversion. This direct benefit transfer system reduces the grip of intermediaries who have historically exploited farmers by siphoning off entitlements or charging informal fees.

Beyond immediate monetary support, such schemes enhance financial inclusion, give farmers greater autonomy in decision-making, and create a sense of dignity by recognising them as direct beneficiaries of state support. In the long run, this contributes to social equity and a more resilient rural economy, where resources are more evenly distributed and smallholders are less vulnerable to exploitation.

Therefore, the future of digital agriculture must be guided by ethical safeguards and transparent governance. Policymakers, technologists, researchers, and farmer organisations need to collaborate in designing systems that are farmer-centric, accountable, and inclusive. This entails establishing clear rules for data ownership, ensuring farmers have control over how their information is utilised, and developing AI tools that are transparent, fair, and culturally relevant. Only by combining innovation with equity, transparency, and farmer empowerment can India ensure that digital transformation becomes a pathway not just to higher productivity, but also to justice, resilience, and shared prosperity in rural life.

6.6 Ethical framework for AI integration

The integration of artificial intelligence (AI) into agriculture holds immense promise for transforming the sector. AI-powered tools can enhance productivity, optimise resource use, improve pest and disease management, and facilitate data-driven decision-making. However, as these technologies become more prevalent, it is essential to develop a robust ethical framework to guide their responsible and fair deployment. An ethical framework ensures that AI benefits all stakeholders—especially vulnerable farmers—

while safeguarding ethical principles such as fairness, transparency, privacy, and accountability.

Without doubt, AI in agriculture has the potential to revolutionise traditional practices, but it also presents risks, including data privacy violations, algorithmic bias, unequal access, and conflicts of interest. Many smallholder farmers, particularly in developing countries like India, may lack the digital literacy or resources to navigate these new technological landscapes, risking marginalisation. Therefore, a comprehensive ethical framework is necessary to prevent misuse of technology, promote inclusive growth, and ensure that AI serves societal and environmental goals. We have identified five core principles of an ethical framework.

6.6.1 Fairness and Inclusivity

AI systems should be designed to benefit all farmers equally, including marginalised and smallholder farmers. This involves ensuring that AI algorithms are trained on diverse data sets to avoid biases and that access to AI-driven tools is affordable and equitable. For example, systems should not favour large-scale commercial farms at the expense of smallholders or disadvantaged communities.

6.6.2 Transparency and Explainability

Farmers and stakeholders must understand how AI tools arrive at recommendations or decisions. Clear communication about data sources, algorithms, and limitations fosters trust and accountability. For instance, if an AI system advises the use of pesticides, farmers should understand the basis of this advice and any associated risks.

6.6.3 Data Privacy and Security

Protecting farmers' data rights is paramount. An ethical framework must establish guidelines for informed consent, data ownership, and secure handling of sensitive information. Farmers should have control over their data and be aware of how it is used and shared.

6.6.4 Accountability and Governance

Developers, governments, and organisations deploying AI must be accountable for the outcomes. Robust monitoring mechanisms should be in place to identify and rectify biases, errors, or harmful impacts. Establishing legal and ethical accountability structures ensures responsibility for unintended consequences.

6.6.5 Environmental Sustainability

AI should promote sustainable farming practices that conserve resources, reduce environmental impact, and support biodiversity. Ethical AI use aligns technological innovation with ecological responsibility.

An ethical AI framework in agriculture must be built on five key principles: fairness and inclusivity, ensuring equal benefits for all farmers, especially smallholders; transparency and explainability, so farmers understand how AI decisions are made; data privacy and security, protecting farmers' rights and ownership of their information; accountability and governance, holding developers, governments, and organisations responsible for outcomes; and environmental sustainability, aligning AI applications with resource conservation, biodiversity, and ecological health. Together, these principles ensure that AI in agriculture is equitable, trustworthy, and sustainable.

6.7 Toward an Ethical AI Framework in Agriculture

Developing an ethical AI framework in agriculture is not just a technical challenge but a societal imperative. It demands multi-stakeholder collaboration, bringing together government agencies, technologists, farmers, researchers, and civil society to co-create systems that are fair, transparent, and accountable. Such a framework must include policies and standards that govern the development, deployment, and monitoring of AI tools. Regular audits, independent oversight bodies, and participatory feedback mechanisms are crucial to ensuring continuous improvement and preventing misuse. In doing so, India—and indeed the world—can harness the full potential of AI while upholding human rights, advancing social justice, and safeguarding environmental health.

India is uniquely positioned to lead this effort. A civilisation rooted in 5,000 years of agricultural practice, holistic thinking, and community-centred living provides fertile ground for shaping an ethical AI paradigm. Indian traditions emphasise harmony with nature, respect for life, and social responsibility—values that resonate deeply with the challenges of modern technology governance. These cultural principles can serve as a moral compass for developing AI systems that empower farmers, mitigate inequities, and safeguard ecosystems.

India has already laid the foundation for ethical, inclusive technology through pioneering initiatives. The Digital Public Infrastructure (DPI)—which includes Aadhaar for identity, UPI for payments, and DigiLocker for secure documentation—has demonstrated how technology can be scaled to millions while remaining affordable and inclusive. In agriculture, initiatives like ITC's e-Choupal have connected farmers directly with markets, reducing exploitation by middlemen and providing access to real-time information. Similarly, AI-driven platforms under the Digital Agriculture Mission are now helping predict crop yields, manage pests, and optimise water use.

AI is also being integrated into crop insurance schemes, such as the Pradhan Mantri Fasal Bima Yojana, where satellite imagery and remote sensing reduce delays and ensure timely compensation to farmers. Pilot projects using AI for drought prediction, soil health monitoring, and climate-smart agriculture are helping smallholders adapt to climate change. These examples demonstrate how technology, when guided by ethical frameworks, can safeguard livelihoods, enhance food security, and foster resilience.

As the world grapples with the ethical dilemmas of emerging technologies, India can offer a model of stewardship rooted in its civilisational ethos and practical innovations. By demonstrating that AI can be designed and governed with societal well-being and ecological health at the forefront, India could set global benchmarks for responsible innovation. This leadership would not only reinforce India's role in shaping the future of agriculture but also establish it as a pioneer in building AI frameworks that advance sustainable progress for humanity as a whole. By uniting its ancient wisdom with modern science, India can show the world that the true power of AI lies not only in what it can do, but in how responsibly it is used to serve people, communities, and the planet.

Summary

This chapter emphasises the importance of a moral compass in guiding decision-making and actions. It underscores that every innovation, regardless of its attractiveness or profitability, should be evaluated for its social impact, particularly on the poor and marginalised communities. Leading agricultural companies now consider people and the environment as integral to their bottom line, alongside profits. Core values such as integrity and transparency are explored through real-world examples, highlighting their role in ethical leadership. With the inevitability of AI integration, the chapter proposes a framework to manage conflicts and eliminate biases, ensuring responsible and socially conscious adoption of technology in agriculture.

Academic Exercise

1. What is meant by the term "moral compass"?
2. Name three value-driven Indian leaders in agriculture.
3. Name three companies following the triple bottom line approach.
4. Describe the term AI-bias and its implications for smallholders.
5. Enumerate five principles of an ethical AI framework in agriculture.

AI

7

Harnessing Artificial Intelligence

AI in agriculture is not a replacement for wisdom, but a bridge—connecting age-old knowledge with modern science to secure a resilient and prosperous future.

7.1 The Value of Local Knowledge

Agriculture is the backbone of many economies, especially in rural and developing regions. At its heart lies the collective wisdom and experience of farmers, passed down through generations, shaped by the unique climate, soil, and social contexts of each locality. Farmers use inherited, local, and observational knowledge to select suitable crops, predict weather, and manage pests with available materials, a practice known as Indigenous Technical Knowledge (ITK). Local knowledge encompasses the skills, practices, and insights that farmers accumulate through firsthand experience with their land, crops, and weather patterns. This wisdom is invaluable.

Farmers determine the best crop varieties for their soil by observing and managing its properties. Farmers assess the texture, colour, and depth of the soil, which helps them identify suitable crops. For instance, clayey soil, with its high water retention, is perfect for water-intensive crops like rice. Alluvial soils, common in river basins, are highly fertile and ideal for various crops, including wheat and sugarcane. To enhance soil health naturally, farmers use traditional methods such as composting crop residues and animal manure, along with spreading nutrient-rich mud from water bodies. Applying sophisticated systems: In some regions, ancient agricultural systems, such as terraced farming, allow for the cultivation of high-value crops like apples on mountainsides while preventing soil erosion.

To predict rainfall patterns, farmers have long observed environmental and biological indicators. Farmers monitor indicators like the direction and strength of winds, the formation and colour of clouds, and atmospheric conditions. For example, dark, thick clouds can signal heavy rain, while grey clouds indicate strong winds. The behaviour of wildlife can also serve as a forecast. In some areas, dragonflies flying close to the ground are considered a sign of impending rainfall. The nesting locations of birds, such as kingfishers, can be used to predict the quantity of seasonal rainfall. The movement of leeches in water can indicate upcoming dry or rainy weather. Observations of celestial bodies, such as the position of the moon crescent and stars, are also used to forecast rain. In India, some farmers consult the Panchang, an ancient almanack, to predict the weather.

For generations, farmers have managed pests using natural and mechanical techniques that prioritise the local ecosystem. Farmers create botanical sprays from local plants to repel pests. Neem oil is a natural insect repellent. Sprays made from garlic and chillies can deter pests like armyworms and caterpillars. In parts of Ethiopia, cow urine is used to protect seedlings from fungi and pests.

Creating natural barriers and traps: Placing sticky substances on tree trunks can prevent crawling insects from reaching the canopy. Traps baited with molasses or other attractants can be used to capture pests like cutworms.

Crop rotation disrupts the life cycle of pests, preventing them from building up in the soil. Companion planting, such as growing onions or garlic alongside other crops, can naturally deter pests. Encouraging beneficial insects, such as ladybugs and parasitic wasps, can provide a natural form of pest control. Using ash for protection: Spreading wood ash in vegetable fields is a traditional method for controlling aphids.

Yet, the rapid pace of technological advancement, shifting market demands, and environmental challenges—such as climate change—have made traditional knowledge alone insufficient for farmers to thrive. Artificial Intelligence (AI) emerges as a powerful tool to bridge the gap between time-honoured local expertise and cutting-edge innovations, fostering efficiency, resilience, and prosperity for farming communities worldwide.

7.2 AI as the Farmer's Knowledge Ally

Agriculture is as much about knowledge as it is about soil, water, and seed. Across generations, farmers have preserved a living library of wisdom—about rains, pests, soils, and survival—that rarely finds its way into formal systems. Artificial Intelligence (AI) offers an unprecedented opportunity to unlock, amplify, and democratise this knowledge, ensuring that it not only survives but also thrives in the service of future generations.

AI does not replace farmers; it empowers them. By capturing local practices through voice, text, and images, and translating them into structured intelligence using natural language processing

(NLP), AI can transform everyday farmer observations into actionable insights. A simple statement—"my paddy leaves are turning yellow after the rains"—becomes a pathway to diagnosing nutrient deficiency, pest damage, or waterlogging. In this way, the informal becomes formal, and the local becomes universal.

At scale, AI-powered platforms can weave individual voices into collective knowledge commons. Multilingual apps, chatbots like Kisan e-Mitra, and digital repositories of peer-vetted practices enable farmers to access, share, and refine solutions. This creates a cycle of learning, sharing, and adapting that strengthens resilience against climate change, pests, and market shocks. Farmers evolve from being passive recipients of advice to active contributors to innovation.

But technology must remain ethical and farmer-centric. Farmers' data should belong to them, algorithms must be transparent, and advisory content should be trusted and locally relevant. Without safeguards, AI risks becoming another instrument of control; with them, it becomes an instrument of liberation. The vision is clear: an agriculture where tradition meets technology, where farmers' wisdom fuels innovation, and where AI becomes a trusted ally in building resilient, just, and sustainable food systems.

7.3 AI acts like a "knowledge weaver"

AI can connect individual farmers to larger networks by transforming local experiences into collective intelligence that benefits entire regions. For instance, when farmers upload pictures or descriptions of sick plants through mobile applications, AI-powered image recognition and natural language processing can diagnose crop diseases quickly. As multiple farmers across villages report similar symptoms, the system aggregates this knowledge to detect emerging outbreaks, issue alerts, and share solutions widely, creating a shared disease surveillance network that protects communities beyond the individual farm.

Similarly, AI-driven soil and climate advisory platforms integrate local farmer data—such as rainfall observations, soil moisture readings, or cropping practices—with satellite imagery and weather forecasts. By aggregating this information, AI generates localised yet scalable recommendations. For example, when several farmers in semi-arid regions record signs of water stress, the system can suggest drought-tolerant crop varieties or optimised irrigation schedules, ensuring that the collective know-how strengthens resilience across the region.

In the economic sphere, AI supports farmers by analysing reports on yields, sale prices, and demand trends. These systems identify patterns and connect smallholders to broader market networks. Suppose farmers in different villages notice falling tomato prices. In that case, AI can aggregate these inputs to predict a regional glut and recommend alternative markets or crop diversification strategies, ultimately improving farmers' bargaining power and market resilience.

AI also facilitates farmer-to-farmer knowledge exchange by matching individuals with similar problems to peers who have already solved them. A farmer in Andhra Pradesh struggling with pest control in groundnuts, for example, could be connected with a counterpart in Karnataka who successfully adopted bio-pesticides. In this way, AI curates and circulates local success stories, enabling the spread of practical, field-tested solutions across regions.

Finally, when farmers report shortages of inputs such as fertilisers or irrigation challenges, AI aggregates these signals and presents them to cooperatives, extension services, and policymakers. This allows for more precise allocation of resources and targeted subsidies, aligning regional agricultural planning with real, ground-level needs. By weaving together countless small threads of experience, AI turns fragmented local knowledge into a robust regional or national network, strengthening both individual farmers and agricultural systems as a whole.

7.4 Enabling Efficiency

AI's ability to synthesise vast amounts of data and generate actionable insights makes it an invaluable asset for improving agricultural efficiency. Artificial Intelligence (AI) can improve agricultural efficiency by synthesising vast amounts of data into actionable insights, helping farmers make smarter decisions that increase yields and reduce costs. By integrating with modern technologies like IoT sensors, drones, and satellite imagery, AI elevates farming from a practice of intuition to one of data-driven precision.

7.4.1 Precision farming and resource optimisation

AI enables precision agriculture, which tailors farming practices to specific field conditions, optimising resources and minimising environmental impact. AI algorithms analyse data from soil moisture sensors and weather forecasts to determine optimal watering schedules. This can reduce water usage by up to 50% compared to traditional, less efficient methods. By analysing soil health data, AI can precisely calculate the ideal amount of nutrients needed in specific areas of a field. This reduces fertiliser waste and minimises runoff, which protects soil and water quality. Using computer vision, AI systems can distinguish weeds from crops and target them with herbicides or robotic tools, leading to up to a 90% reduction in pesticide usage. This targeted approach protects crops, beneficial insects, and the environment.

7.4.2 Predictive analytics and risk management

AI's ability to forecast outcomes allows farmers to manage risk proactively and adapt to changing conditions. AI analyses historical and real-time weather data to provide more accurate local forecasts. This helps farmers plan planting, irrigation, and harvesting schedules more effectively, reducing losses from adverse weather. By synthesising data from past seasons, soil conditions, and weather patterns, AI models can forecast crop yields with high accuracy. This helps farmers plan for logistics, storage, and market demands. AI-powered image recognition can analyse images from drones or field cameras to detect signs of disease or nutrient deficiencies long before they are visible to the human eye. This allows for timely intervention and prevents widespread crop damage.

7.4.3 Automation and efficiency

AI-driven automation reduces labour needs, increases operational efficiency, and allows for more consistent performance. Robots and self-driving tractors guided by AI can perform labour-intensive tasks like planting, weeding, and harvesting with high precision. This helps farmers combat labour shortages and reduces overall operational costs. AI algorithms analyse crop readiness and weather conditions to determine the ideal time for harvest, minimising waste and maximising the crop's market value. AI-powered computer vision systems can accurately sort produce by quality, size, and ripeness, which is far more efficient and accurate than manual sorting methods.

7.4.4 Empowering data-driven decisions

For both small and large-scale farms, AI provides the data and tools needed to make smarter, more profitable decisions. AI-powered software integrates data from various sources to provide a comprehensive overview of farm operations. This helps farmers optimise resource allocation, manage budgets, and enhance overall profitability. AI can forecast market demand, optimise transportation logistics, and improve traceability from farm to consumer. This reduces food waste and increases efficiency across the entire supply chain.

7.5 Fostering Prosperity

Efficiency is only one dimension of prosperity. AI can help farmers achieve holistic well-being by promoting inclusivity, resilience, and sustainable growth. Through tailored, accessible tools, AI can address multifaceted challenges and create a more equitable, empowered, and prosperous agricultural ecosystem.

7.5.1 Promoting inclusivity and equity

AI can democratise access to vital agricultural information and financial services, which is particularly beneficial for smallholder and marginalised farmers. AI-powered multilingual chatbots and advisory services can deliver real-time, hyperlocal information on crop management, soil health, and market prices in local languages. This bridges communication gaps for farmers who lack access to formal extension services or digital literacy, ensuring they can make data-driven decisions. Traditional lending often excludes smallholder farmers due to a lack of formal credit history. AI models can use alternative data—like satellite imagery showing crop health or mobile phone usage patterns—to create comprehensive credit profiles. This helps financial institutions assess risk more accurately, enabling loans for seeds, equipment, and other investments. AI-driven platforms can connect small-scale farmers directly with buyers, reducing dependency on intermediaries and ensuring better prices for their produce. For example, the Saagu Baagu project in Telangana, India, used AI to create a digital marketplace for chilli farmers, significantly increasing their income.

7.5.2 Building resilience to environmental and economic shocks

AI can help farmers proactively adapt to climate change and market volatility, reducing stress and providing greater stability. AI models can analyse weather patterns, soil conditions, and historical data to provide more accurate, localised forecasts. This allows farmers to make informed decisions about when to plant, irrigate, or harvest, mitigating risks from droughts, floods, or pest outbreaks. AI also accelerates the breeding of climate-resilient crops by analysing large genetic datasets. By analysing market trends and historical data, AI can predict commodity prices and demand fluctuations. This helps farmers plan their crop choices and sales timing to maximise profits and minimise losses from market shifts. AI can facilitate data-sharing platforms among farming communities. This network enables farmers to share real-time information about production threats, such as pests or diseases, allowing for coordinated responses that are more effective than individual actions.

7.5.3 Promoting sustainable and regenerative growth

Beyond simply optimising resource use, AI can support farming practices that regenerate ecosystems and foster long-term prosperity. AI analyses soil microbiome data to provide insights into soil health and fertility. This helps farmers optimise regenerative practices, such as crop rotation and cover cropping, which enrich the soil, improve water retention, and enhance carbon sequestration. By enabling precision agriculture, AI helps farmers reduce their reliance on chemical fertilisers and pesticides. The Saagu Baagu project showed a 9% reduction in pesticide use and a 5% decrease in fertiliser usage, resulting in lower costs and a minimised environmental impact. AI-powered irrigation and nutrient management systems ensure the precise application of resources, leading to significant reductions in water and input waste. This not only lowers costs for the farmer but also protects natural resources for future generations.

7.6 Real-world Examples

Artificial Intelligence (AI) is steadily transforming Indian agriculture, offering solutions to some of the sector's most persistent challenges—low productivity, unpredictable climate patterns, resource inefficiency, and fragile market linkages. What makes this revolution particularly significant is its focus on inclusivity, with many AI-powered tools being designed for smallholder farmers who form the backbone of Indian agriculture. By working in local languages and running on affordable mobile devices, these innovations are bridging the digital divide, ensuring that advanced technology is not confined to large agribusinesses but reaches the last mile in rural villages.

7.6.1 Pest and Disease Detection – Plantix

The Plantix mobile app has emerged as one of the most widely used AI tools among Indian farmers. Developed by PEAT (Progressive Environmental & Agricultural Technologies), it allows farmers to upload photos of their crops and instantly receive a diagnosis. The app uses AI-based image recognition to identify more than 350 plant diseases, pests, and nutrient deficiencies. Beyond diagnosis, it provides actionable recommendations in local languages, helping farmers take timely preventive and corrective measures. By digitising and aggregating millions of farmer inputs, Plantix also builds a collective database of pest and disease outbreaks, which can be used for early warning systems across regions.

7.6.2 Weather Forecasting and Climate Adaptation – IBM Watson

In collaboration with the Government of India, IBM Watson Decision Platform for Agriculture provides hyper-local weather forecasts and personalised advisories to farmers. Using AI, IoT, and satellite data, the platform predicts rainfall patterns, temperature fluctuations, and pest outbreaks with high precision. Farmers receive this information as SMS alerts or through mobile apps in regional languages. For example, a farmer in Maharashtra might be warned about an upcoming dry spell, enabling them to adjust irrigation schedules or delay sowing. By combining climate science with AI, this initiative helps farmers mitigate risks posed by climate change, reducing crop losses and improving resilience.

7.6.3 Crop Monitoring and Yield Prediction – Cropin

The Bengaluru-based startup Cropin leverages AI, satellite imagery, and big data to help farmers and agribusinesses monitor crops, predict yields, and assess risks. Its platform, "SmartFarm," enables farmers to record detailed farm-level data, which is then processed by AI models to track crop growth and forecast yields. Banks and insurance companies also use Cropin's data to make lending and insurance decisions, reducing risk for both farmers and financial institutions. In states like Madhya Pradesh and Uttar Pradesh, Cropin has helped digitise thousands of farms, improving transparency in the agricultural value chain. This system ensures farmers receive targeted advisory support while enabling governments and corporates to make better resource allocations.

7.6.4 Soil and Irrigation Management—Microsoft & ICRISAT Collaboration

Microsoft, in collaboration with the International Crops Research Institute for the Semi-Arid Tropics (ICRISAT), developed an AI-powered sowing advisory system. By analysing soil data, historical weather records, and real-time forecasts, the system recommends optimal sowing dates to farmers in Andhra Pradesh and Karnataka. The AI tool also suggests irrigation schedules and fertiliser use based on soil moisture levels and predicted rainfall. Farmers who adopted the recommendations saw yield improvements of up to 30%, showing the power of AI to enhance traditional farming practices. Importantly, the service was designed for accessibility—it works on basic feature phones and delivers advisories via SMS in local languages, ensuring inclusivity for smallholder farmers.

7.6.5 Market Linkages – Ninjacart

Ninjacart, one of India's leading agritech startups, uses AI and predictive analytics to connect farmers directly with retailers, restaurants, and wholesalers. Traditionally, farmers have had to rely on multiple intermediaries, often receiving low prices for their produce. Ninjacart's AI-driven platform forecasts demand, optimises logistics, and ensures that produce reaches markets faster and fresher. Farmers benefit from better prices and reduced wastage, while buyers get more reliable supplies. For example, a tomato farmer in Karnataka can sell directly to urban retailers in Bengaluru without going through multiple middlemen. By streamlining the supply chain with AI, Ninjacart has significantly reduced post-harvest losses, which are a major issue in Indian agriculture.

7.6.6 Farm Mechanisation – Hello Tractor Partnership

Mechanisation is often out of reach for smallholder farmers due to the high costs of owning tractors and equipment. To address this, Mahindra & Mahindra, India's largest tractor manufacturer, partnered with Hello Tractor, an Africa-based Agri-tech startup, to launch an AI-enabled tractor rental platform in India. Farmers can request tractors and equipment through a mobile app or SMS, utilising a model similar to an "Uber for tractors." AI algorithms optimise tractor deployment, ensuring that machines are allocated efficiently across villages based on demand. This has made modern mechanisation accessible and affordable for smallholders, helping them increase efficiency during critical sowing and harvesting periods.

Autonomous tractors would enable a farmer to focus on the work that matters the most on a farm.

Anand Mahindra

The growing adoption of AI in Indian agriculture demonstrates how technology, when combined with local knowledge and inclusive design, can drive systemic transformation. By helping farmers diagnose crop issues, adapt to climate variability, optimise resources, and access fairer markets, AI is not only improving productivity but also enhancing resilience and livelihoods. Importantly, these initiatives highlight the need for continued collaboration among farmers, innovators, and policymakers to ensure that benefits reach even the most marginalised communities. As India moves toward a future of sustainable and technology-driven farming, AI will serve as both an enabler and a bridge—linking traditional wisdom with modern science to secure food security and rural prosperity.

7.7 A Collaborative Vision

The future of AI in agriculture will be shaped not by technology alone, but by the depth of collaboration among farmers, researchers, technologists, and policymakers. Farmers hold the most valuable asset—practical, context-rich knowledge of local soils, crops, and weather patterns. However, this knowledge is often fragmented and unstructured. Researchers play a crucial role in translating these field-level insights into scientific frameworks, validating them, and ensuring that AI models are grounded in agronomic realities rather than abstract datasets. Technologists, in turn, are responsible for building the tools and platforms that make AI accessible, scalable, and adaptable to the diversity of agricultural systems—from smallholder farms to large-scale agribusiness. Without user-friendly interfaces, multilingual NLP systems, and cost-effective sensors, AI solutions would remain out of reach for the majority of farmers.

Policymakers act as bridges and enablers, creating the regulatory frameworks, incentives, and infrastructure needed for AI to thrive in agriculture. This includes investing in digital connectivity in rural areas, setting standards for data privacy, and ensuring that smallholders are not excluded from the benefits of emerging technologies. Moreover, policies that support open data sharing, farmer cooperatives, and public–private partnerships will ensure that AI applications serve the public good rather than only commercial interests.

When these four groups collaborate effectively, AI can evolve into a participatory and inclusive enterprise. Farmers address real-world problems and provide feedback; researchers provide evidence-based validation; technologists design adaptive solutions; and policymakers create enabling ecosystems. Together, they can ensure that AI addresses pressing challenges such as climate change, food security, and sustainable resource use, while also empowering rural communities economically. In this sense, the future of AI in agriculture is not just technological, but social and collaborative, driven by collective intelligence and shared responsibility.

Regulatory bodies and governments create the policies, laws, and infrastructure necessary for AI to develop responsibly. They invest in rural connectivity and digital literacy to ensure equitable access, while setting standards for data privacy, ownership, and transparency to protect farmers' rights. At the same time, they act as funders and facilitators of innovation, channelling resources into crops and regions that may not attract private investment but are vital for food security. Through incentives, regulations, and international coordination, policymakers align AI adoption with broader goals of sustainability, equity, and resilience. Together, this collaboration ensures that AI in agriculture is not just a technological advance but a participatory and inclusive enterprise. Farmers contribute ground-level insights, researchers provide scientific validation, technologists design adaptive solutions, and policymakers establish enabling frameworks.

In the end, the true power of AI will not lie in algorithms, but in alliances. When human wisdom converges with artificial intelligence, and when collaboration replaces competition, a new paradigm of agriculture will emerge—one that feeds both people and the planet. The digital field of the future will not be a battlefield of profit, but a commons of shared purpose, where data becomes dialogue, and every farmer becomes a co-creator of knowledge. This is the promise of a collaborative intelligence—rooted in empathy, guided by science, and bound by our shared destiny on this living Earth.

Summary

This chapter highlights AI's dual nature as both an opportunity and a risk, while underscoring its transformative role in agriculture. AI serves as a catalyst and enabler by capturing agricultural knowledge, driving precision agriculture, risk management, automation, and data-driven decisions. It enhances efficiency and builds resilience against environmental and economic shocks, with real-world examples illustrating its impact. AI emerges as a collaborative and generative force, capable of reshaping agriculture into a more sustainable and innovative enterprise. AI in agriculture is not just a technology, but a bridge between traditional wisdom and modern science, securing resilience, productivity, and prosperity for farmers.

Academic Exercise

1. Enumerate traditional knowledge in agriculture that withstood the test of time.
2. Why is AI necessary to curate the traditional knowledge?
3. Explain AI as an efficiency enabler.
4. How does AI help build resilience to environmental and economic shocks?
5. Articulate AI as a collaborative vision.

8

Learning to Lead

Inclusive leadership in agriculture grows where mentorship plants the seed, leadership tends the soil, and technology nurtures the harvest.

8.1 Social Inclusion as a Core Principle

Agriculture has long been the backbone of human civilisation, shaping cultures, economies, and environments. Yet, in the face of climate change, widening inequalities, and rising food insecurity, conventional farming models have proven insufficient. Holistic and inclusive agriculture emerges as a transformative approach that not only prioritises productivity but also ensures ecological sustainability, social justice, and economic resilience. It recognises that farming is not just about growing crops—it is about nurturing ecosystems, empowering communities, and safeguarding the future of food.

Modern industrial farming systems, while highly productive, have often relied on chemical inputs, monocropping, and unsustainable water use. These practices have degraded soils, eroded biodiversity, and created vulnerabilities to climate shocks. A holistic approach to agriculture views the farm as part of a living ecosystem, emphasising harmony between soil, water, crops, livestock, and people. Techniques such as agroecology, organic farming, permaculture, and regenerative practices enhance soil fertility, conserve water, and restore ecological balance. By valuing natural cycles, holistic agriculture reduces dependency on external inputs and builds resilience against environmental uncertainties.

Agriculture is not just an economic activity; it is a social system involving farmers, labourers, and consumers. Historically, marginalised groups such as women, smallholders, indigenous communities, and landless labourers have often been excluded from agricultural benefits.

Social justice is not a charity or political slogan—it is a fundamental right. A nation's progress depends on including its most vulnerable in the journey of development.

Dr B. R. Ambedkar

Inclusive agriculture ensures their active participation and fair share in the agricultural value chain. Women-led self-help groups in Kerala, for instance, have demonstrated how collective farming can enhance livelihoods and promote social empowerment. Youth engagement in agriculture, supported by technology and entrepreneurship programs, also ensures generational continuity. Recognising indigenous knowledge systems—such as traditional cropping methods in tribal regions—further strengthens cultural sustainability and inclusiveness. Agriculture needs leaders who cultivate not just the land, but the confidence, skills, and dreams of those who work on it.

One of the central challenges for small-scale farmers is limited access to fair markets, which often leads to exploitation by middlemen. Inclusive agriculture emphasises economic equity through farmer cooperatives, farmer-producer organisations (FPOs), and community-supported agriculture (CSA) models. These structures enable farmers to pool resources, negotiate collectively, and access more favourable markets. In India, the promotion of FPOs has given small farmers a stronger voice in pricing, procurement, and supply chain participation. Moreover, inclusive agriculture encourages diversification of income sources—through allied activities like beekeeping, poultry, agroforestry, and food processing—thus reducing dependence on a single crop and increasing resilience to market shocks.

The integration of digital technologies in agriculture—ranging from AI-driven crop monitoring to IoT-enabled precision farming—has immense potential. However, technology must be inclusive. Many small farmers lack access to smartphones, stable internet, or digital literacy. Holistic and inclusive agriculture insists that technological advances be democratised, made affordable, and adapted to local contexts. For instance, mobile-based advisory platforms in India provide farmers with real-time weather forecasts, pest alerts, and market price updates in regional languages. When combined with community training programs, these tools ensure that innovation does not widen inequalities but instead uplifts the most vulnerable.

8.2 Real-World Examples

Governments and regulatory bodies play a crucial role in developing inclusive agricultural systems that empower farmers and support sustainable communities. Beyond regulation, they must create enabling environments where smallholders, women, and marginalised groups can thrive by ensuring access to subsidies, credit, insurance, and sustainable technologies. Inclusive policies should protect the dignity of farm labour through fair wages, safe conditions, and gender equity, while also promoting cooperatives and community ownership to strengthen resilience. By linking farmers to institutional buyers and public procurement, governments not only guarantee income stability but also reinforce food security. In doing so, state action becomes a catalyst for fairness, resilience, and sustainability, aligning economic growth with social justice.

8.2.1 Andhra Pradesh's Zero-Budget Natural Farming (ZBNF)

Andhra Pradesh's Zero-Budget Natural Farming (ZBNF) is a groundbreaking initiative that seeks to transform agriculture by eliminating farmers' dependence on costly chemical inputs and replacing them with natural, locally available resources. Introduced in 2016 under the Rythu Sadhikara Samstha (RySS), the program aims to transition six million farmers across eight million hectares to a sustainable, climate-resilient model of farming.

The core practices of ZBNF—such as Beejamrutha (seed treatment), Jeevamrutha (soil enrichment using cow dung, cow urine, jaggery, and pulse flour), mulching to conserve soil moisture, and Whapasa (maintaining soil aeration rather than water saturation)—help restore soil fertility, reduce water usage by up to 40%, and enhance biodiversity. By drastically cutting input costs, ZBNF improves farmers' profitability, diversifies cropping systems, and strengthens resilience against climate shocks. Socially, it empowers women farmers through self-help groups and promotes farmer-to-farmer learning networks, making the program both participatory and inclusive.

Globally recognised by organisations like the FAO, ZBNF is hailed as the world's largest agroecology program. However, it faces challenges such as yield uncertainties during the transition phase, scientific debates over its scalability, and the need for strong policy and market support. Despite these hurdles, Andhra Pradesh's ZBNF represents a pioneering movement toward ecological balance, economic sustainability, and social empowerment, offering a viable alternative to chemical-intensive agriculture and a potential model for global adoption.

8.2.2 Kerala's Kudumbashree Movement

Kerala's Kudumbashree Movement is one of India's most successful women-led poverty eradication and empowerment initiatives, launched in 1998 by the Government of Kerala in collaboration with the National Bank for Agriculture and Rural Development (NABARD). Built around a vast network of women's self-help groups (SHGs), Kudumbashree empowers women by providing access to microfinance, livelihood opportunities, and social development programs. In agriculture, it has been especially transformative: women's collectives lease land, cultivate crops jointly, and adopt sustainable practices that not only enhance food security but also provide steady incomes for rural households. The movement promotes inclusive growth by bringing marginalised women into the economic mainstream, strengthening their bargaining power, and fostering leadership at the grassroots level. Beyond farming, Kudumbashree extends into micro-enterprises, skill training, health, and social welfare, making it a holistic development model. Today, it encompasses millions of women across Kerala, serving as a powerful example of how community-driven initiatives can combine social empowerment, economic resilience, and collective action to build equitable and sustainable rural economies.

8.2.3 Two Different Approaches

Kerala's Kudumbashree Movement and Andhra Pradesh's Zero-Budget Natural Farming (ZBNF) represent two distinct yet complementary models of people-centred agricultural transformation in India. While Kudumbashree emphasises women's empowerment through collective action, enabling self-help groups to lease land, cultivate crops, and generate income, ZBNF focuses on ecological sustainability and cost reduction by promoting natural farming practices that eliminate dependence on chemical inputs. Kudumbashree addresses social and economic inclusion by mobilising marginalised women into decision-making and livelihoods, whereas ZBNF tackles environmental and financial challenges faced by farmers by restoring soil health, conserving water, and reducing production costs. Both models strengthen community resilience and local food systems, but their approaches differ: Kudumbashree builds empowerment from a social and institutional perspective, while ZBNF drives change through an agroecological and climate-resilient lens. Together, they demonstrate how inclusive policies and grassroots participation can transform agriculture from a mere means of survival to a pathway to equity, sustainability, and dignity for rural communities.

8.3 Developing AI Skills

Developing AI skills in agriculture leaders is becoming increasingly essential as farming faces new challenges in sustainability, climate resilience, and productivity. Artificial intelligence has the potential to revolutionise how crops are cultivated, monitored, and marketed, but its effectiveness depends on whether agricultural leaders—policymakers, researchers, cooperative heads, and farmer organisation leaders—can understand and apply these tools responsibly. For this, leaders must build strong foundations in data literacy, learning how to interpret outputs from satellites, sensors, and predictive models, while also developing the ability to make informed decisions using AI-generated insights. Beyond technical expertise, leaders need to foster digital transformation within their communities, guiding cooperatives and institutions to adopt AI in ways that remain accessible to small and marginal farmers. Ethical awareness is equally important, ensuring AI is transparent, unbiased, and inclusive.

Training agricultural leaders in AI requires a blend of structured learning and practical exposure. Universities and government institutions can provide specialised courses on machine learning, geospatial analytics, and digital agriculture, while experiential learning through pilot projects and exposure visits can help leaders apply these concepts in real-world contexts. Collaborative platforms that bring together technologists, agribusinesses, and farmer organisations can also facilitate peer-to-peer learning and innovation. Blended models that combine online theoretical training with offline workshops for hands-on practice are particularly effective. In addition, leaders must learn how to translate complex AI insights into farmer-friendly communication, ensuring that knowledge reaches the grassroots in local languages and actionable formats.

Despite its promise, the path to AI-enabled leadership in agriculture presents challenges. The rural digital divide remains a significant barrier, with many areas lacking reliable internet connectivity to support AI applications. Leaders may also face resistance from farmers who are wary of unfamiliar technologies, highlighting the need for building trust through the demonstration of reliable results. High costs of advanced AI tools and a lack of technical skills can slow adoption, while ethical concerns around farmer data use demand careful governance. These obstacles underline the importance of building AI capacity in ways that are affordable, accessible, and socially inclusive.

Several global examples illustrate the importance of leadership in AI adoption for agriculture. In India, platforms like CropIn and AgNext demonstrate how trained leaders can use AI tools for precision farming and quality assessment, improving farmer adoption. In Kenya, AI-powered dairy apps help cooperatives monitor cattle health and increase milk yields, showing how leadership drives impact. The Netherlands' use of AI in smart greenhouses highlights how visionary leadership can integrate high-tech solutions into sustainable farming, while Andhra Pradesh has used data-driven monitoring to support natural farming practices under its Zero-Budget Natural Farming program. Each case shows that leaders trained in AI act as crucial bridges between technology and communities.

Looking ahead, the way forward lies in creating structured opportunities for leadership development in AI. Governments can establish AI leadership academies, universities can design agrifood-AI curricula, and private sector partnerships can sponsor fellowships or exchange programs. Importantly, training must be rooted in local contexts, ensuring that AI complements traditional knowledge rather than displacing it. By fostering leaders who are fluent in both agricultural realities and technological innovation, the sector can evolve into a climate-smart, equitable, and tech-enabled system. Ultimately, developing AI skills in agriculture leaders is not just about embracing technology—it is about building inclusive, resilient, and sustainable pathways for the future of farming.

8.4 AI literacy and interdisciplinary thinking

AI literacy is the ability to understand, interpret, and apply artificial intelligence tools in ways that are meaningful and practical for decision-making. In agriculture, this means that leaders and practitioners should not only know how AI systems function but also be able to evaluate the insights these systems produce. For example, predictive analytics may indicate the likelihood of pest outbreaks, shifts in rainfall patterns, or fluctuations in market prices. A leader with AI literacy can interpret such results, assess their reliability, and translate them into actionable strategies for farmers. This skill set also includes understanding data quality, recognising the limitations of AI models, and ensuring ethical use of farmer data.

Interdisciplinary thinking complements AI literacy by allowing leaders to connect knowledge from multiple domains—technology, ecology, economics, and social sciences—to solve complex agricultural challenges. For instance, deploying an AI-based irrigation system requires not just technical knowledge of sensors and algorithms but also an understanding of soil characteristics, water availability, farmer behaviour, and local governance structures. Similarly, AI-driven crop forecasting is more effective when integrated with insights from climate science, market analysis, and traditional farmer wisdom. Interdisciplinary thinking also helps balance innovation with inclusivity, ensuring that solutions remain grounded in local realities rather than being imposed as one-size-fits-all technologies. Leaders who think across disciplines can bring together agronomists, data scientists, economists, and community organisations to co-create solutions that are scientifically robust, economically feasible, environmentally sustainable, and socially just. In this way, interdisciplinary thinking ensures that AI is not used in isolation but as part of a broader system of knowledge that addresses agriculture's diverse challenges.

8.5 AI for farmer education

AI for farmer education begins with the ability to personalise learning. Unlike conventional "one-size-fits-all" training models, AI systems can analyse farmer profiles, local climate data, soil conditions, and cropping patterns to deliver tailored advice. For instance, an AI-powered mobile app can notify a farmer about the best time to sow seeds, detect early signs of crop disease from a photo, or suggest water-saving practices based on weather forecasts. By providing knowledge that is specific, relevant, and accessible, AI not only enhances productivity but also builds farmer confidence in adopting sustainable practices.

Another powerful dimension of AI in farmer education is its capacity for language inclusion. Many farmers, particularly in countries like India, operate in regional languages and dialects, making conventional training materials less effective. Natural Language Processing (NLP), a branch of AI, enables the translation of complex agricultural information into simple, local-language formats. Interactive voice assistants and chatbots, for example, can answer farmers' questions in their mother tongue, breaking down barriers of literacy and technical jargon. This makes education more inclusive, ensuring that even farmers with limited formal schooling can benefit from digital knowledge.

AI also plays a crucial role in scaling agricultural extension services. Human extension workers, while important, cannot always reach every village or farm. AI-driven platforms, however, can serve millions of farmers simultaneously, delivering real-time updates on weather, pest outbreaks, and market prices. During critical times, such as droughts or pest infestations, AI-based early warning systems can disseminate life-saving information at scale, equipping farmers with strategies to mitigate losses. In this way, AI amplifies the reach and efficiency of extension networks, making them more responsive and resilient.

Beyond technical knowledge, AI contributes to holistic farmer education by linking farmers to markets, financial services, and peer networks. Platforms powered by AI can teach farmers about market trends, contract farming opportunities, and digital payments, thus broadening their economic literacy alongside agricultural skills. Virtual communities, facilitated by AI algorithms, connect farmers across regions to share experiences and local innovations, fostering peer-to-peer learning. Such integration makes farmer education not just about crop yields but about building resilience, entrepreneurship, and long-term sustainability.

However, the use of AI for farmer education is not without challenges. The digital divide—lack of internet connectivity, limited smartphone access, and low digital literacy—remains a major barrier in rural areas. There is also the risk of over-reliance on AI tools without adequate human support, which could lead to mistrust if recommendations fail. Moreover, ethical concerns such as data privacy and algorithmic bias must be addressed to ensure fairness and transparency. These limitations highlight the need for blended approaches, where AI complements rather than replaces traditional extension systems. Training extension workers to use

AI tools, investing in rural connectivity, and co-creating solutions with farmer communities are essential steps for making AI-driven education truly inclusive.

AI holds immense promise in reimagining farmer education by making knowledge more personalised, inclusive, and scalable. From detecting pests and predicting weather to providing financial literacy and market insights, AI can equip farmers with the skills and confidence to thrive in a rapidly changing agricultural landscape. Yet, its success depends on ensuring accessibility, trust, and ethical use. By combining the strengths of technology with human extension systems and local knowledge, AI can evolve into a powerful enabler of farmer education—one that empowers not only individual farmers but also entire rural communities to embrace sustainable and resilient futures.

8.6 AI in extension services

Agricultural extension services have traditionally played a central role in bridging the gap between research institutions and farmers by providing technical advice, training, and innovations. However, conventional extension systems are often constrained by limited personnel, high costs, and difficulties in reaching dispersed farming communities. With the rapid growth of digital technology, artificial intelligence (AI) has begun to reshape extension services, making them more accessible, efficient, and responsive to farmers' needs.

Unlike traditional extension models that often provide generalised advice, AI enables precision outreach, improving both productivity and resource efficiency.-Without doubt, AI has the potential to revolutionise extension services by making them more personalised, inclusive, and scalable. When combined with human expertise and localised knowledge, AI can enhance the capacity of extension workers, empower farmers with timely and accurate information, and foster resilience in the face of climate and market uncertainties. The future of extension services lies in a hybrid model where artificial intelligence supports, rather than replaces, human-centred extension systems, ensuring that technology becomes an enabler of sustainable and equitable agricultural development.

8.7 Mentorship and leadership programs

Mentorship and leadership programs are vital in shaping the next generation of agricultural innovators, policymakers, and community leaders. In a sector undergoing rapid transformation due to climate change, globalisation, and digital technologies, such programs provide structured opportunities for young farmers, researchers, and agri-entrepreneurs to gain guidance, confidence, and vision. Mentorship connects emerging professionals with experienced practitioners who can share practical insights, technical expertise, and lessons learned from the field. This relationship not only accelerates skill development but also builds networks of trust and collaboration that are essential for sustainable agricultural growth.

Leadership programs go a step further by equipping individuals with the ability to inspire, coordinate, and drive collective action within farming communities and institutions. These programs often focus on critical skills such as strategic thinking, communication, negotiation, and policy advocacy, preparing participants to tackle complex agricultural challenges. By integrating modules on sustainability, AI literacy, climate-smart practices, and social inclusion, leadership programs ensure that future agricultural leaders are both technologically competent and socially responsible. For example, farmer-producer organisations (FPOs) in India increasingly rely on trained leaders who can manage resources, engage with markets, and advocate for farmers' rights at the policy level.

Importantly, mentorship and leadership initiatives promote inclusivity by empowering women, youth, and marginalised groups to take on decision-making roles in agriculture. Women-led self-help groups in Kerala's Kudumbashree or youth-focused agri-entrepreneurship programs demonstrate how targeted mentorship can unlock hidden potential within rural communities. When combined with exposure visits, peer-learning circles, and international exchange programs, these initiatives cultivate global perspectives while keeping leaders grounded in local realities.

Mentorship and leadership programs serve as catalysts for agricultural transformation, creating a cadre of well-informed, confident, and visionary individuals who can integrate traditional knowledge with modern innovations. By fostering both personal growth and community resilience, such programs ensure that agriculture remains not only a source of livelihood but also a driver of equitable, sustainable, and future-ready development. The best leaders in agriculture are not those who grow the most crops, but those who grow the most people. Mentorship is the seed, leadership is the soil, and together they grow the next generation of sustainable farmers.

Summary

This chapter emphasises social inclusion as the foundation of agricultural leadership, illustrating this through initiatives such as Zero Budget Natural Farming (ZBNF) in Andhra Pradesh, which enhances ecological and economic resilience, and Kudumbashree in Kerala, which advances women's empowerment and collective livelihood development. It argues that future leaders must cultivate artificial intelligence (AI) skills and interdisciplinary thinking, as digital tools have the capacity to transform farmer education, strengthen extension services, and create data-driven, networked approaches to agricultural development. The chapter concludes by affirming the centrality of mentorship in leadership formation, encapsulated in the metaphor that mentorship is the seed, leadership is the soil, and together they nurture the next generation of sustainable farmers.

Academic Exercise

1. What is the core principle of agricultural leadership?
2. Why is collective livelihood development crucial in India?
3. What is interdisciplinary thinking?
4. What is the paramount role AI plays in extension services?
5. Explain the metaphor about mentorship and leadership.

9

Practical Pathways

> ***The true power of AI in agriculture is not in making us faster or smarter, but in helping us act wisely—building equity, resilience, and sustainability from vision to reality.***

9.1 From Ideas to Action

Ideas are the seeds of change, but they remain inert unless nurtured into action. Moving from the realm of imagination to the realm of reality requires vision, courage, planning, and persistence. History shows that progress has always been shaped by individuals and communities who refused to stop at "what could be" and instead pushed toward "what is possible."

Every great transformation begins as a simple idea—whether it is a scientific theory, a social reform, or a business innovation. Ideas give us direction and reveal alternatives to existing problems. Yet, ideas alone cannot feed the hungry, reduce inequality, or build technology. They are powerful only when translated into action. The journey from thought to execution is often obstructed by fear, inertia, and lack of resources. Many ideas fade away because of hesitation or resistance to change. Society sometimes undervalues dreamers, labelling their ideas as "impractical". This is why leadership, collaboration, and persistence are critical in pushing ideas beyond the conceptual stage.

Moving from ideas to action requires a series of deliberate steps that transform imagination into tangible outcomes. The first step is clarity and purpose. An idea, no matter how brilliant, will remain abstract unless it is refined into a clear vision. Defining the purpose involves asking what problem the idea seeks to solve, who will benefit from it, and what change it aims to bring. Without this clarity, an idea often loses direction and fades away before it can take shape.

The next step is planning and strategy. Once the vision is defined, it must be supported with a realistic and practical plan. Planning provides structure by setting clear goals, identifying resources, allocating time, and anticipating challenges. A well-designed strategy ensures that passion and creativity are guided into productive action rather than scattered efforts. It provides a roadmap that helps the dreamer stay focused, even when obstacles arise.

Execution and adaptability form the third stage. Implementing an idea is often the most demanding part of the journey. It requires discipline, consistency, and resilience. Rarely does a plan succeed on the first attempt, and this is where adaptability becomes crucial. Setbacks and mistakes should not be viewed as failures, but rather as opportunities to learn and refine one's approach. Those who adapt and persevere are the ones who eventually succeed in converting ideas into reality.

Equally important is collaboration and support. No idea grows in isolation. Family, mentors, peers, and communities provide encouragement, guidance, and resources that an individual may not have access to alone. Collaboration also allows the idea to be tested, challenged, and improved through diverse perspectives. Support networks help sustain momentum and scale ideas from small beginnings into larger movements or enterprises.

Finally, evaluation and growth complete the cycle. Once an idea has been acted upon, it is necessary to reflect on the journey. Evaluating what worked and what needs improvement not only strengthens the original idea but also provides valuable lessons for future endeavours. Growth comes from this continuous cycle of reflection, adaptation, and innovation, ensuring that ideas do not just stop at one achievement but continue to inspire further action.

One individual may die for an idea, but that idea will, after his death, incarnate itself in a thousand lives.

Netaji Subhas Chandra Bose

In today's world, artificial intelligence, renewable energy, and digital platforms are proof that ideas can rapidly reshape industries and economies. What was unimaginable a decade ago is now a reality because innovators turned inspiration into implementation. The lesson is clear: action multiplies the value of an idea. Ideas are sparks of human creativity, but only through action do they ignite meaningful change. The future belongs not just to dreamers but to doers—those who are willing to step into uncertainty, experiment, and persist. As the saying goes, "Vision without action is merely a dream; action without vision is merely passing time; but vision with action can change the world."

9.2 Roadmaps for aspiring future leaders

Agriculture today stands at a crossroads where traditional practices meet modern challenges such as climate change, food insecurity, and market volatility. For aspiring future leaders in agriculture, a clear roadmap is essential to guide their journey from learning to leading. Leadership in this sector requires not only knowledge of farming practices but also vision, adaptability, and the ability to inspire communities toward sustainable growth.

The first step in this roadmap is building a strong foundation of knowledge. Future leaders must gain a deep understanding of both conventional farming practices and modern innovations. Knowledge of soil health, water management, and crop diversity must be complemented with exposure to emerging tools such as precision agriculture, artificial intelligence, and data-driven decision-making. This balance of tradition and technology ensures that leaders are well-prepared to address diverse agricultural challenges.

The second step is developing problem-solving and interdisciplinary thinking skills. Agriculture is not an isolated field—it is connected to economics, ecology, technology, and social sciences. Aspiring leaders must learn to view problems from multiple perspectives and design holistic solutions. For example, tackling declining soil fertility requires not just technical fixes but also policy reforms, community engagement, and financial planning. Interdisciplinary thinking makes agricultural leadership innovative and future-oriented.

The third step involves social inclusion and community engagement. True leaders in agriculture must place farmers and rural communities at the heart of their vision. They should learn from grassroots models like Andhra Pradesh's Zero Budget

Natural Farming (ZBNF) or Kerala's Kudumbashree movement, which highlight empowerment, inclusivity, and self-reliance. By encouraging participation from women, youth, and marginalised groups, leaders can build resilient agricultural systems that benefit all.

The fourth step in the roadmap is embracing mentorship and networking. Leadership cannot be developed in isolation. Aspiring leaders must seek guidance from experienced farmers, researchers, and policymakers, while also creating peer networks for mutual learning. Mentorship provides wisdom, while networking creates opportunities for collaboration and scaling impactful ideas.

The final step is cultivating a vision for sustainability and resilience. Future agricultural leaders must look beyond short-term productivity and prioritise long-term ecological balance. Climate-smart practices, renewable energy, and circular economies must be central to their plans. Leaders should see themselves not only as food producers but also as stewards of the environment and guardians of food security for future generations. The roadmap for aspiring leaders of agriculture combines knowledge, interdisciplinary thinking, social inclusion, mentorship, and sustainability. By following this path, they will not only strengthen their own leadership journey but also contribute to building resilient food systems that can nourish humanity while protecting the planet. The future of agriculture will not be written only in fields and farms, but in the vision of leaders who can sow innovation, harvest inclusion, and cultivate sustainability.

9.3 AI for Funding

Agriculture is the backbone of rural livelihoods, yet one of its persistent challenges is access to timely and fair funding. Small and marginal farmers, cooperatives, and Agri-startups often struggle to secure loans or investments because of limited collateral, weak credit histories, and bureaucratic hurdles. Artificial Intelligence (AI) has the potential to revolutionise agricultural funding by making financial systems more inclusive, efficient, and responsive to farmers' real needs.

One of the most promising uses of AI in agriculture is in credit assessment. Traditional banks evaluate farmers mainly through land records or repayment history, excluding millions who lack formal documentation. AI can analyse alternative data sources such as crop yield patterns, weather conditions, mobile payment transactions, and even satellite imagery to create accurate and fair credit scores. This allows deserving farmers to access loans without being penalised for lack of collateral.

AI can also streamline loan approvals and reduce delays. Automated systems can quickly review applications, predict repayment capacity, and identify risks. For farmers, this means faster access to funds during critical agricultural seasons such as sowing and harvesting, when delays can severely affect productivity. By reducing human bias and paperwork, AI ensures a smoother, farmer-friendly funding process.

Additionally, AI-powered platforms can serve as matchmakers between farmers and funding opportunities. These systems can connect farmers, cooperatives, or Agri-startups with investors, government subsidies, or microfinance institutions that align with their goals. For example, a farmer adopting organic farming methods could be automatically linked with sustainability-focused investors or climate-resilient agriculture funds.

Another vital contribution of AI is in risk management and insurance. By analysing weather forecasts, soil data, and crop health patterns, AI can predict potential risks, such as droughts or pest attacks. This helps funding agencies design tailored loan products and insurance schemes that protect farmers from debt traps. Moreover, AI can detect fraudulent claims and misuse of funds, thereby building trust between lenders and borrowers.

Most importantly, AI can promote financial inclusion and equity in rural areas. It can highlight underfunded regions, support women farmers, and identify marginalised groups that are often overlooked by traditional finance. Policymakers can then use these insights to design targeted subsidies and schemes that bridge funding gaps.

Without doubt, AI can play a transformative role in agricultural funding by making it more accessible, transparent, and secure. By integrating technology into credit scoring, loan approval, matchmaking, and risk management, AI ensures that farmers are not left behind in the financial system. When harnessed wisely, it empowers farmers not only with resources but also with confidence to innovate, invest, and lead agriculture into a sustainable future.

9.4 Harnessing Tech Startups

Agriculture has always been the foundation of human survival. Still, in today's rapidly changing world, it faces complex challenges such as climate change, shrinking landholdings, and the growing demand for food security. Traditional practices alone are no longer sufficient to meet these demands. This is where technology-driven startups have emerged as game changers, bringing innovation, efficiency, and sustainability into agriculture. Harnessing the potential of these startups can transform farming into a more productive, profitable, and future-ready sector.

One of the most significant contributions of Agri-tech startups is in data-driven farming. Startups are using artificial intelligence, drones, sensors, and satellite imagery to help farmers monitor soil health, water usage, and crop growth in real time. This precision agriculture ensures that inputs such as seeds, fertilisers, and water are used optimally, reducing costs and increasing yields. Farmers who once relied on guesswork can now make informed decisions based on accurate data.

Tech startups are also playing a crucial role in improving market access. For decades, farmers have suffered losses due to middlemen and poor connectivity with buyers. Digital platforms created by startups now allow farmers to sell directly to consumers, retailers, or food processing industries. By ensuring transparency in pricing and creating digital marketplaces, startups empower farmers to secure fair prices for their produce. Another area where startups are making an impact is financial inclusion and risk management. Many farmers struggle to access loans and insurance due to a lack of records or collateral. Startups are building AI-powered credit scoring systems, mobile-based microfinance services, and customised insurance products. These innovations make funding accessible to small-scale farmers and protect them from risks such as crop failure or extreme weather.

Sustainability is another vital domain where Agri-tech startups are contributing. From promoting organic farming and hydroponics to creating solutions for efficient water use and renewable energy, startups are aligning agriculture with environmental goals. These eco-friendly innovations not only improve resilience against climate change but also appeal to conscious consumers who demand healthier, sustainable food. Perhaps most importantly, Agri-tech startups are fostering youth engagement in agriculture. By blending farming with technology, they make the sector more attractive to younger generations who see it not as a backward occupation but as a field of innovation and entrepreneurship. This shift is crucial for ensuring the long-term vitality of the agricultural sector.

Harnessing Agri-tech startups for agriculture holds the promise of solving some of the sector's toughest challenges while opening new avenues for growth. By supporting these innovators through policy reforms, investments, and collaborations, agriculture can evolve into a modern, technology-powered enterprise. Startups are not just supporting farmers; they are reshaping the future of food security, rural development, and environmental sustainability.

One of the most direct ways AI can empower smallholders is through precision farming. By analysing data from weather forecasts, soil sensors, and satellite imagery, AI can give farmers real-time advice on when to plant, irrigate, and harvest.

9.5 AI for Sustainable Innovation

Artificial Intelligence (AI) has emerged as one of the most powerful tools of the 21st century, transforming industries and redefining human potential. Beyond efficiency and productivity, AI carries immense promise for sustainability, enabling societies to innovate in ways that reduce environmental harm, optimise resources, and build resilience. By integrating AI into sustainable practices, we can address global challenges such as climate change, food security, and energy conservation while creating inclusive pathways for growth.

One of the most impactful roles of AI in sustainable innovation is resource optimisation. AI systems can analyse vast amounts of data to minimise waste and maximise efficiency in sectors such as agriculture, water management, and energy. For instance, precision agriculture powered by AI ensures that seeds, fertilisers, and water are used only where necessary, reducing environmental stress while improving yields. Similarly, AI-driven smart grids in energy systems balance demand and supply, enabling efficient use of renewable sources like solar and wind.

AI agents will become the primary means of interaction with computers in the future. They will be able to understand our needs and preferences, and proactively help us with tasks and decision-making.

Satya Nadella

Integrating AI for smallholder empowerment holds immense potential to transform the agriculture sector into a more inclusive and resilient one. By improving productivity, managing risks, expanding financial access, and enhancing market connectivity, AI equips small farmers with the tools they need to thrive. When used responsibly and ethically, AI can shift smallholders from a position of vulnerability to one of strength, ensuring that they remain central players in feeding the world.

AI also supports climate action and environmental monitoring. Satellite-based AI models can track deforestation, soil erosion, and biodiversity loss with high accuracy, providing real-time data for policymakers and conservationists. Predictive analytics can forecast extreme weather events, enabling communities to prepare and adapt accordingly. By transforming data into actionable insights, AI enhances global efforts to protect ecosystems and mitigate climate-related risks.

Another dimension of sustainable innovation lies in circular economy models. AI can streamline waste management by predicting consumption patterns, improving recycling systems, and designing smarter supply chains. By identifying inefficiencies and promoting reuse, AI contributes to reducing the linear "take-make-dispose" model of production, moving industries toward sustainable cycles.

In addition, AI can empower social sustainability by promoting inclusivity. Platforms powered by AI can give smallholder farmers, micro-entrepreneurs, and marginalised communities access to knowledge, markets, and financial services. This not only drives innovation at the grassroots level but also ensures that sustainable development benefits all sections of society rather than a privileged few.

One of the most important ways AI promotes inclusivity is by bridging the information gap. Many communities, particularly in rural areas, lack access to timely knowledge about education, healthcare, or agriculture. AI-powered platforms, equipped with translation tools and voice assistants, can deliver critical information in local languages and even through audio for those with low literacy levels. This ensures that people who were previously excluded from mainstream information channels can now participate fully in social and economic opportunities.

AI also contributes to inclusivity by expanding access to essential services. For example, in healthcare, AI chatbots and diagnostic tools are bringing medical advice to remote areas where doctors are scarce. In education, AI-driven adaptive learning platforms personalise content for students from diverse backgrounds, allowing children with different learning abilities to thrive. By making services more accessible, AI helps reduce inequalities between urban and rural communities, as well as between wealthy and poor households.

AI also has the potential to amplify the voices of underrepresented individuals. Natural Language Processing (NLP) tools can translate diverse dialects, capture grassroots inputs, and connect marginalised groups to policymakers. This ensures that development policies reflect the realities of people on the ground rather than only those of privileged groups. Inclusivity here is not just about access but about participation and representation.

Despite these advantages, it is equally important to address ethical and governance concerns. AI must be developed and deployed responsibly, ensuring transparency, fairness, and accountability. If misused, it risks widening inequality or reinforcing unsustainable consumption. Sustainable innovation through AI must therefore be guided by human values, policies, and global cooperation.

AI is not just a tool for efficiency but a catalyst for sustainable innovation. By optimising resources, enabling climate action, advancing the circular economy, and empowering communities, AI has the potential to reshape the path of development. When used wisely, it bridges the gap between technological progress and environmental stewardship, ensuring a future that is both innovative and sustainable. Artificial Intelligence should not just make us faster and smarter—it must help us live lighter on the planet and fairer with one another.

Summary

This chapter emphasises how ideas must move beyond vision to action, offering a roadmap for aspiring agricultural leaders to create real impact. It highlights how Artificial Intelligence can strengthen access to funding by connecting farmers with financial institutions, assessing risks, and ensuring resources are used effectively. Agri-tech startups are presented as powerful partners, capable of delivering innovative platforms and services that provide farmers with timely information, market access, and practical solutions tailored to their needs. Together, these pathways encourage leadership that is both forward-looking and inclusive.

The chapter also explores AI's potential to empower farmers directly, from optimising resources to building resilience and strengthening participation in agricultural value chains. Beyond immediate gains, AI is framed as a tool for sustainable innovation—reducing environmental impact, conserving resources, and promoting fairness across communities. The conclusion underscores a broader vision: Artificial Intelligence should not merely make humanity faster and smarter, but help us live more lightly on the planet and more equitably with one another.

Academic Exercise

1. Chart a course of an idea becoming an action.
2. What are the most promising uses of AI in agriculture?
3. Which is the most direct way AI can empower smallholders?
4. What is a hallmark of data-driven farming?
5. Why is AI not just a tool for efficiency but a catalyst for sustainable innovation?

10

The Future We Want

The future of food is not about producing more, but about producing with purpose—where AI amplifies human wisdom to nourish people, protect the planet, and empower farmers.

10.1 Modern Food System

The modern food system is one of the most complex networks in human history, connecting farmers, processors, distributors, retailers, and consumers worldwide. Unlike traditional food systems, which were largely local and subsistence-based, today's system operates on a global scale, shaped by technological innovation, market integration, and evolving consumer demand. This interconnected web ensures that fresh produce, processed foods, and diverse cuisines are available year-round, even in regions where they are not locally produced. However, such abundance also comes with challenges related to sustainability, equity, and health.

One defining feature of the modern food system is its efficiency and scale. Mechanised farming, biotechnology, digital supply chains, and global trade networks have dramatically increased food production and distribution. These advancements have reduced hunger in many regions, diversified diets, and enabled rapid responses to changing market needs. At the same time, the system is heavily influenced by multinational corporations and industrial farming practices, which prioritise profit and productivity. While this has lowered costs and increased availability, it has also contributed to ecological concerns such as soil degradation, biodiversity loss, and greenhouse gas emissions.

The health dimension of the modern food system is equally significant. Processed and ultra-processed foods dominate supermarket shelves, offering convenience and affordability but also fuelling rising rates of obesity, diabetes, and other non-communicable diseases. The paradox of abundance coexisting with malnutrition—where undernourishment and overconsumption occur simultaneously—reveals deep inequalities embedded in the system. Moreover, smallholder farmers and marginalised communities often struggle to compete, despite being essential to local food security. Even though they form the backbone of food production in many parts of the world, they remain some of the most vulnerable actors within the modern food system.

Despite their critical role in sustaining local food security, they often face structural disadvantages such as limited access to credit, modern technology, reliable markets, and fair pricing mechanisms. Global supply chains tend to favour large-scale agribusinesses that can produce at scale and negotiate better terms, leaving smallholders squeezed between rising production costs and volatile market prices. In addition, marginalised communities frequently lack adequate infrastructure—such as storage facilities, transportation, and irrigation systems—which further weakens their competitiveness. Climate change adds another layer of hardship, as unpredictable weather patterns, droughts, and floods disproportionately affect those with the fewest resources to adapt. These struggles highlight a paradox: the very communities that grow much of the world's food often lack food security themselves. Addressing this imbalance requires inclusive policies, targeted support, and innovative technologies that empower smallholders, ensuring they are not left behind in the rapidly evolving modern food system.

Looking forward, the modern food system must balance efficiency with responsibility. Innovations such as precision agriculture, plant-based alternatives, artificial intelligence in supply chains, and sustainable packaging offer pathways toward resilience and equity. A truly modern food system should not only feed the world but also safeguard ecosystems, protect farmer livelihoods, and ensure that nutritious food is accessible to all. Ultimately, the future of food depends on transforming the system from one that merely maximises production to one that sustains both people and the planet. A truly modern food system is not measured by how much it produces, but by how fairly it nourishes people and how gently it treads upon the planet.

10.2 Food Systems that Empower and Sustain

Food is more than nourishment—it is the foundation of health, livelihoods, and culture. In an era where industrial agriculture dominates, reimagining food systems to both empower people and sustain the planet has become an urgent priority. A food system that empowers is one that places farmers, especially smallholders and marginalised communities, at the centre of decision-making. Empowerment comes through equitable access to land, credit, technology, markets, and education. When farmers are empowered, they are not passive suppliers to a global chain but active stewards of food security and agents of innovation. Such systems strengthen communities by improving incomes, encouraging cooperation, and preserving local traditions of cultivation and consumption.

Equally vital is the sustainability of food systems. Current models often prioritise efficiency and yield at the expense of soil health, biodiversity, and climate stability. A sustainable food system, however, ensures that natural resources are used wisely, ecosystems are protected, and future generations can thrive. Approaches such as agroecology, regenerative farming, and climate-smart agriculture demonstrate that productivity and sustainability need not be opposing goals. By integrating technology, such as artificial intelligence for precision resource use, alongside traditional knowledge, societies can create resilient systems that adapt to environmental pressures while minimising ecological footprints.

A shining example of leadership in creating food systems that empower and sustain is the legacy of M.S. Swaminathan, the father of India's Green Revolution. His vision combined science with social responsibility, ensuring that technological breakthroughs in high-yielding crop varieties reached small farmers and transformed India from a food-deficient nation into a food-secure one. Importantly, Swaminathan's philosophy went beyond mere production—he consistently emphasised the need for an "Ever-Green Revolution", one that increased productivity without harming the environment. By championing farmer-centric policies, gender inclusion in agriculture, and sustainable innovations, he embodied the principle that true food security must nourish people, empower communities, and protect the planet.

Another towering figure in this vision was Verghese Kurien, celebrated as the Father of the White Revolution. Through Operation

Flood, he transformed India into the world's largest producer of milk, turning scarcity into abundance. Kurien's genius lay not only in organising supply chains but in empowering rural communities through cooperative models. By placing dairy farmers—especially women—at the heart of production, he ensured fair prices, decentralised decision-making, and community ownership. His legacy proves that food systems rooted in inclusion and fairness can both sustain livelihoods and strengthen national resilience.

Equally inspiring is the story of Shri Mahila Griha Udyog Lijjat Papad, a women-led cooperative that began in Mumbai with just seven women and a borrowed loan of eighty rupees. What started as a small home-based venture grew into a nationwide movement employing thousands of women, giving them financial independence, dignity, and a voice in society. Though modest in scale compared to industrial agribusiness, Lijjat Papad symbolises how collective action, rooted in inclusion and self-reliance, can empower marginalised groups while sustaining cultural food traditions. It stands as a reminder that food systems are not only about production but also about cooperation, fairness, and community solidarity.

The same spirit of empowerment is embodied by the Self-Employed Women's Association (SEWA), which has organised millions of women workers across India, including small farmers, vendors, and food processors. By securing access to credit, markets, and training, SEWA has enabled rural women to become not just participants but leaders in local economies. Its model emphasises dignity of labour, cooperative strength, and self-reliance, proving that inclusion is central to sustainable food systems. SEWA's work underscores that when the most marginalised are empowered, entire communities gain resilience.

Ultimately, food systems that empower and sustain must balance social equity with ecological responsibility. They must ensure that nutritious food is affordable and accessible to all, while also recognising and rewarding the efforts of those who grow and process it. In doing so, they move beyond the narrow aim of feeding the world toward the higher goal of nourishing people, communities, and the planet in harmony.

"The true measure of a food system is not how much it produces, but how deeply it empowers people and how gently it sustains the Earth."

10.3 Combining human ingenuity with advanced AI

Agriculture has always been a field where human ingenuity shines—farmers adapting to shifting climates, communities preserving seeds across generations, and innovators creating tools to make cultivation more efficient. Today, artificial intelligence offers a new set of possibilities, from precision farming to predictive analytics. Yet, the true transformation emerges not from AI alone, but from its combination with human creativity and wisdom. Farmers understand the rhythms of soil, water, and seasons in ways that machines cannot. When this local knowledge is paired with AI's ability to process vast data—such as weather forecasts, soil health indicators, and market trends—agriculture becomes both more productive and more resilient.

For instance, AI-powered platforms can suggest drought-resistant crops for semi-arid regions, but it is the farmer's ingenuity that adapts these recommendations by cultivating traditional millet varieties that fit local diets and cultural practices. Similarly, soil sensors can provide precise recommendations for the quantity of nitrogen or potassium to apply; however, farmers often combine these insights with organic compost or manure, ensuring that precision does not come at the expense of ecological balance. In pest management, AI can detect early outbreaks using data from drones or satellites, but farmers often combine this with traditional methods, such as intercropping marigolds or adjusting irrigation patterns to deter pests. Even in market decisions, while AI may forecast price fluctuations, farmers sometimes choose to reserve part of their harvest for barter or community needs, reflecting social priorities no algorithm can capture.

The role of ISRO (Indian Space Research Organisation) further demonstrates how human ingenuity and advanced AI intersect to enhance agricultural productivity. ISRO's satellite imagery and remote sensing data, when analysed through AI algorithms, provide precise insights into soil moisture, crop health, and water availability. These applications support farmers in making better decisions about irrigation, crop selection, and disaster preparedness. For example, AI models trained on ISRO's datasets are being used for drought early-warning systems, enabling timely interventions in vulnerable regions. Similarly, the Bhuvan geoportal, enhanced by AI, helps communities and policymakers monitor land use and manage resources efficiently. Yet, it is the ingenuity of farmers, extension workers, and local leaders that translates these high-tech solutions into practical strategies—turning space science into tools for local resilience.

This partnership between human ingenuity and AI is especially vital for smallholder farmers, who often face constraints in resources, infrastructure, and access. AI-powered tools can guide them, but it is their creativity and cultural wisdom that determine how these technologies are used in diverse landscapes. By aligning AI with the values of sustainability and equity, agriculture can move beyond its narrow focus on higher yields toward the broader goal of nourishing communities while protecting ecosystems.

The future of farming will depend on how effectively we harness AI to extend human capabilities—empowering farmers, conserving the planet, and ensuring that technology serves not just efficiency but also fairness and sustainability. The future of agriculture will not be written by algorithms alone, but by the wisdom of farmers empowered through artificial intelligence. AI may power the tools of tomorrow, but it is the wisdom of farmers that will shape their purpose—together forging a future where technology serves not just productivity, but fairness, sustainability, and the nourishment of life itself.

10.4 Cultivating leadership rooted in ethics, innovation, and inclusion

Agriculture is not only the backbone of food security but also a sector where leadership deeply influences communities, ecosystems, and economies. Cultivating leadership rooted in ethics ensures that decisions about land, water, and technology are guided by fairness, transparency, and responsibility. Ethical agricultural leaders recognise that farming is not simply about yields, but about safeguarding soil health, protecting biodiversity, and ensuring that farmers—especially smallholders—are treated with dignity. Ethics in agriculture means resisting exploitative practices, advocating for fair markets, and making choices that sustain both present and future generations.

Yet ethics alone are not enough. The agricultural sector demands innovation to adapt to climate change, resource scarcity, and growing population needs. From AI-driven precision farming and digital marketplaces to regenerative practices and climate-smart techniques, innovation provides tools that increase productivity while reducing environmental harm. However, innovation must be shaped by inclusion to be truly transformative. Inclusive agricultural leadership ensures that women farmers, rural youth, and marginalised communities are active participants in change, not passive recipients of it. By democratising access to new knowledge, credit, and technologies, inclusive leaders create systems where every farmer—regardless of scale—can thrive.

A powerful example of such leadership is reflected in the work of R. S. Paroda, former Director General of ICAR, who spearheaded the creation of India's National Gene Bank. His initiative was rooted in the ethical responsibility to safeguard the genetic diversity of crops, ensuring that future generations would not lose access to the very seeds of life. By preserving indigenous varieties and wild relatives, he demonstrated foresight and innovation, recognising that resilience in agriculture begins with biodiversity. At the same time, his vision was deeply inclusive: the gene bank was not meant to serve only scientific communities, but also farmers, who could access conserved genetic resources to strengthen their livelihoods. Paroda's leadership illustrates how ethics, innovation, and inclusion can converge—protecting heritage, embracing modern science, and empowering farmers to build sustainable futures.

Equally inspiring is the contribution of Dr A. P. J. Abdul Kalam, who, though remembered globally as India's "Missile Man" and former President, always kept farmers and rural communities close to his heart. Through his vision of PURA (Providing Urban Amenities in Rural Areas), articulated with Prof. P. V. Indiresan, former Director of the Indian Institute of Technology (IIT), Madras, Dr Kalam championed the idea that smallholder farmers and rural youth must have access to the same infrastructure, technology, and opportunities as their urban counterparts. PURA combined ethics—uplifting the most disadvantaged—with innovation—leveraging science, ICT, and renewable energy—and inclusion—ensuring rural communities were active participants in India's growth story. In doing so, Kalam showed that true leadership bridges science with compassion and technology with equity.

When ethics, innovation, and inclusion come together in agricultural leadership, they lay the foundation for resilient food systems that nourish both people and the planet. Such leadership transcends technical efficiency, instead striving for social justice, ecological stewardship, and shared prosperity. The future of agriculture depends not only on growing more food, but also on cultivating leaders who are ethical in vision, innovative in action, and inclusive in spirit.

10.5 AI Is Inevitable; Better Embrace It

Artificial intelligence is no longer a futuristic concept—it is already transforming agriculture in profound ways. From predictive analytics to smart irrigation systems, AI has become an inevitable force shaping how food is grown, distributed, and consumed. In a world facing climate change, resource constraints, and rising food demand, resisting AI would mean clinging to outdated methods that cannot meet modern challenges. By embracing AI, farmers and policymakers can unlock new levels of efficiency, resilience, and sustainability, ensuring that agriculture remains not only productive but also adaptable to an uncertain future.

The value of AI in agriculture lies in its ability to complement human ingenuity. AI-powered platforms can forecast rainfall, detect early pest infestations, and guide the precise use of water and fertilisers, but it is the farmer's wisdom that adapts these insights to local soil, traditions, and markets. For instance, Indian agri-tech startups like Fasal and AgNext are building AI solutions tailored for crop monitoring, pest detection, and quality assessment—technologies that can be adapted globally. These tools democratise access to information, connect smallholders to markets, and reduce losses, proving that local innovation powered by AI can address global challenges.

The rise of Indian-origin global tech leaders demonstrates why embracing AI is both inevitable and empowering. Satya Nadella at Microsoft has emphasised AI as a tool to enhance productivity and democratise access to knowledge. Sundar Pichai at Google has called AI "more profound than fire or electricity," highlighting its transformative potential across industries, including agriculture. Arvind Krishna at IBM has championed trustworthy AI, pushing for systems that are transparent, ethical, and human-centred. Their leadership shows how India's ethos of resilience and inclusivity is shaping global AI frameworks, reinforcing the message that AI must serve humanity rather than dominate it. Farmers, innovators, and policymakers in India stand to benefit by embracing this wave—guided by leaders who are already setting the global AI agenda.

Smallholder farmers stand to gain the most when AI tools are designed to be affordable, accessible, and inclusive. Yet, this opportunity comes with responsibility: bridging the digital divide, embedding ethical safeguards, and ensuring that no community is left behind. When adopted thoughtfully, AI can elevate agriculture beyond the limited pursuit of higher yields toward the greater purpose of nourishing people and protecting ecosystems. The future of farming will not be defined by a choice between tradition and technology, but by a fusion of both, where human wisdom and digital intelligence work hand in hand to sustain life. The promise of AI in agriculture is not in replacing the farmer's wisdom, but in amplifying it—so that tradition and technology together can feed the future.

10.6 Change starts from Within

At its deepest level, agriculture is far more than the mechanics of tilling land, sowing seeds, and reaping harvests—it is a mirror of the human spirit. The principle that "change starts from within" reminds us that farming is both an external practice of cultivation and an internal journey of self-discovery. Just as the soil must be prepared with care before it can receive seeds, the human mind and heart must be readied for resilience, balance, and stewardship. When a farmer shifts from dependence to self-reliance, from extraction to regeneration, and from fear of scarcity to trust in nature's abundance, the land itself begins to respond. Fields flourish when the farmer's outlook is rooted in patience, humility, and reverence—virtues that cultivate not only crops but continuity of life.

Philosophers have long taught that inner change is the seed of outer renewal. In Andhra Pradesh's Zero-Budget Natural Farming, the choice to embrace natural cycles over synthetic inputs is not simply a technical adjustment; it is a philosophical declaration of trust in the earth's innate intelligence. Likewise, Kerala's Kudumbashree movement shows that when women are empowered from within—nurturing confidence, dignity, and leadership—entire communities are transformed. These are living examples of how inner renewal radiates outward, reshaping both societies and ecosystems.

Agriculture, then, becomes a teacher of life itself. The soil responds to the soul: when human beings cultivate respect, patience, and gratitude within themselves, the earth yields abundantly in return. However, when greed, haste, or exploitation prevail, the land reflects that imbalance through degradation and loss. The lesson is profound: lasting change cannot be imposed from the outside; it must grow from within. By transforming their inner relationship with nature, farmers—and indeed all of humanity—can awaken a renewal that transcends fields and seasons, creating a future rooted in harmony, dignity, and sustainability. The soil of the earth responds to the soil of the soul—when we cultivate within, the harvest outside follows.

10.7 Innovate Locally to Succeed Globally

Agriculture has always been shaped by local knowledge—generations of farmers adapting to their soils, climates, and cultural practices. In today's interconnected world, this local wisdom is not only relevant but also essential to global success. The principle "innovate locally to succeed globally" highlights the idea that true innovation emerges when global challenges are addressed through local solutions. Farmers who experiment with indigenous seeds, natural methods, or region-specific technologies are not merely sustaining their communities; they are pioneering models that the wider world can learn from. Local innovation grounds agriculture in authenticity, resilience, and sustainability—qualities urgently needed in global food systems.

Fresh examples from across India and beyond demonstrate this principle in action. In Sikkim, the decision to become the world's first 100% organic state has provided a local model for chemical-free farming that has inspired international dialogues on sustainable agriculture. In Africa, the widespread adoption of M-Pesa mobile banking has enabled farmers to access credit and digital payments, proving that rural financial innovation can have global relevance. Similarly, Latin America's agroforestry systems show how blending traditional farming with modern ecological science can enhance productivity while conserving biodiversity. These cases highlight how rooted, context-specific practices can travel across borders as global solutions.

Artificial intelligence now offers an opportunity to take local innovations to a global scale. AI-powered tools that analyse soil health, forecast pest outbreaks, or connect smallholders to nearby markets are most effective when tailored to local realities. For instance, Indian startups such as Fasal and AgNext are developing region-specific AI solutions for crop monitoring and quality assessment, which can be adapted and replicated globally. By innovating locally—combining AI with cultural knowledge and ecological practices—farmers and Agri-entrepreneurs can set global benchmarks for sustainable and inclusive food systems. When farmers innovate in their own soil, they plant ideas that can feed the world.

10.8 Living the Legacy

Mahatma Gandhi once envisioned India as a nation of self-reliant village republics, where prosperity flowed not from central power but from empowered communities. His dream was of villages that were independent yet interconnected, rooted in dignity, fairness, and sustainability. While industrialisation and globalisation shifted attention away from this vision, today, artificial intelligence offers a powerful means to bring Gandhi's idea back to life—reshaped for the 21st century. Far from being a technology of distant cities or elite centres, AI has the potential to flow into the very heart of rural India, transforming agriculture into a space of empowerment, innovation, and resilience.

"My idea of village swaraj is that it is a complete republic, independent of its neighbours for its own vital wants, and yet interdependent for many others, which dependence is a necessity."

Mahatma Gandhi

AI is already reshaping the agricultural landscape by democratising knowledge and expertise. Gandhi believed that every farmer should be a custodian of wisdom, not dependent on external exploitation. Today, AI-powered advisory systems like Microsoft's AI Sowing App provide farmers with personalised planting guidance using weather and soil data, while DeHaat and CropIn use AI to deliver crop health diagnostics and market linkages directly to farmers' mobile phones. Just as Gandhi emphasised local empowerment, these innovations return control to the hands of smallholders, reducing dependence on middlemen and creating digital autonomy. By using AI to optimise resources, minimise waste, and connect to fair markets, villages can become hubs of sustainable prosperity—echoing Gandhi's principle of Gram Swaraj (self-rule of villages).

Today, AI can extend this cooperative spirit into the digital era—platforms like AgNext utilise AI-based quality assessment to ensure fairness in milk, grains, and spices, while ISRO's Bhuvan portal, powered by AI, enables villages to collectively manage land and water. These modern tools amplify Kurien's cooperative model, combining community solidarity with cutting-edge technology to build fairer and more resilient food systems.

AI provides the very tools to realise this vision: precision farming that reduces chemical overuse, smart irrigation that conserves water, and data-driven crop planning that balances yield with ecological health. By blending Gandhi's ethical ideals, Kurien's cooperative strength, and Swaminathan's sustainability framework, AI can usher in a farming future that is both productive and compassionate; the idea of Evergreen Revolution will finally be realised.

"If agriculture goes wrong, nothing else will have a chance to go right. The Evergreen Revolution involves improvement of productivity in perpetuity without ecological or social harm."

M. S. Swaminathan

The future of agriculture will not be written in silos of soil and seed alone, but in the convergence of biology, information, and nanotechnology—fused into the intelligence of AI. This is the frontier of an Evergreen Revolution, one that promises not just higher yields but resilient ecosystems, nourished communities, and a planet restored. India, with its legacy of the Green Revolution and its rising technological prowess, stands uniquely positioned to lead this transformation. By marrying the timeless wisdom of the land with cutting-edge science, Indian agriculture can demonstrate to the world how innovation, stewardship, and leadership can together feed humanity while healing the Earth.

Summary

This chapter envisions a transformation of the modern food system, arguing that it must move beyond the narrow aim of producing and trading for profit toward a model that empowers farmers and sustains the planet. It emphasises that agriculture should not simply feed populations but also nourish communities, protect ecosystems, and ensure equity for those who cultivate the land. Central to this transformation is the idea of combining human ingenuity with advanced AI, where technology amplifies rather than replaces the wisdom of farmers. The chapter emphasises the importance of cultivating leadership grounded in ethics, innovation, and inclusion, ensuring that progress in agriculture is driven by responsibility, creativity, and fairness. It asserts that AI is inevitable, and farmers must embrace it as a partner in resilience rather than fear it as a threat.

Academic Exercise

1. How could the modern food system balance efficiency with responsibility?
2. Articulate the role of ISRO in infusing AI in agriculture.
3. Who called for safeguarding the genetic diversity of crops, and how?
4. Which is the world's first 100% organic state?
5. How is Mahatma Gandhi's dream of self-reliant village republics revived?

11

Torchbearers

Civilisations endure not by the harvests they reap, but by the consciousness with which they sow.

11.1 Creators of Lasting Consciousness

Agriculture has always been more than a means of subsistence; it is the foundation of civilisation, the engine of livelihoods, and a mirror of humanity's progress. From the first seeds sown by early farmers to the latest gene-edited crops, agriculture has shaped societies, cultures, and economies. The story of agriculture is not only about soil, seeds, and harvests, but also about visionaries who imagined farming as a civilisational project. These leaders combined science, philosophy, and social consciousness to turn agriculture into a force that could uplift humanity. In this chapter, we honour ten global philosophers and scientists who created lasting consciousness about agriculture—not merely as a science but as a cornerstone of human well-being and progress.

Jethro Tull (1674–1741, England): Mechanisation and Rational Farming

Jethro Tull, often hailed as the father of modern agriculture, revolutionised farming through mechanisation. His invention of the seed drill in 1701 mechanised sowing, ensuring uniform planting and higher yields. Beyond engineering, Tull advocated for systematic soil cultivation, arguing that tillage could replace manure as a means of soil fertility. His ideas sparked debates, but they also established the principle that agriculture must be based on observation, experimentation, and rational improvement—values that remain central to modern agricultural science.

Eva Ekeblad (1724–1786, Sweden): Food Security through Potatoes

Eva Ekeblad, a Swedish countess and agronomist, revolutionised rural livelihoods by domesticating the potato into a staple food. Her innovation of converting potatoes into flour and alcohol eased grain shortages and created new pathways for nutrition security. In an age of famine, her work highlighted how science could solve food crises and empower communities. Ekeblad's legacy is a reminder that the introduction of one resilient crop can change the fate of nations.

Gregor Mendel (1822–1884, Austria): The Father of Genetics

A humble monk in Brno, Mendel conducted meticulous experiments with pea plants and uncovered the laws of inheritance. His discovery of dominant and recessive traits laid the foundation for genetics, a field that would later reshape agriculture through plant breeding, hybridisation, and biotechnology. Mendel exemplifies how quiet curiosity, grounded in patient observation, can ignite revolutions that extend far beyond the monastery garden. His legacy continues to guide scientists in crop improvement and precision breeding.

Erich Tschermak (1871–1962, Austria): Hybrid Vigour in Action

Building upon Mendel's principles, Erich Tschermak reintroduced and applied genetic laws to agriculture. He pioneered hybrid plant breeding, demonstrating how crossing different varieties could produce higher-yielding, more resilient crops. His contributions made hybridisation a practical tool, ushering in new levels of productivity. By translating Mendel's theories into applied science, Tschermak bridged the gap between genetics and the farmer's field, laying the foundation for the hybrid seed industry.

Sir Albert Howard (1873–1947, England/India): Organic Agriculture's Prophet

Albert Howard, often referred to as the father of modern organic farming, developed his ideas in India, where he observed and respected the traditional farming practices of local farmers. He advocated composting, soil health, and biodiversity, stressing that farming should harmonise with natural ecosystems. His seminal book *An Agricultural Testament* inspired the organic movement worldwide. Howard's philosophy—that soil health underpins human health—remains a guiding principle in sustainable agriculture and regenerative farming movements today.

George Washington Carver (1864–1943, USA): Agriculture for Human Dignity

Born into slavery, Carver became one of the most inspirational agricultural scientists of the 20th century. At Tuskegee Institute, he promoted crop diversification, especially peanuts and sweet potatoes, to free farmers from cotton dependency. He linked agriculture to nutrition, soil restoration, and economic independence. For Carver, farming was not just a technical pursuit but a means of human dignity and empowerment, particularly for marginalised communities. His work remains a beacon for linking agricultural science with social justice.

Norman Borlaug (1914–2009, USA): The Green Revolution

Norman Borlaug, Nobel laureate and architect of the Green Revolution, is credited with saving over a billion people from hunger. Through the development of high-yielding, disease-resistant wheat varieties, he transformed food production in Asia and Latin America. Borlaug's pragmatic philosophy—that science must serve humanity—made him a global symbol of agricultural leadership. He demonstrated that technology and policy, when combined, could alleviate hunger on a massive scale.

Yuan Longping (1930–2021, China): The Father of Hybrid Rice

Yuan Longping's breakthrough in developing high-yielding hybrid rice varieties lifted millions out of poverty and hunger in China and beyond. He devoted his life to reducing famine and believed that agriculture was the noblest of human vocations. His vision extended beyond the lab, as he often stated that his dream was to see rice growing taller than himself and providing enough food for all. Yuan exemplified the moral compass of agricultural science—research dedicated to global well-being.

William Dar (b. 1953, Philippines): Champion of Inclusive Agriculture

As former Director General of the International Crops Research Institute for the Semi-Arid Tropics (ICRISAT) and Agriculture

Secretary of the Philippines, William Dar has championed "inclusive market-oriented development". His leadership emphasises climate resilience, smallholder empowerment, and integrating science with policy. Dar represents a modern agricultural leader who recognises that productivity gains must translate into prosperity for farmers, especially in vulnerable regions. His philosophy binds technology with inclusivity.

J. Scott Angle (b. 1953, USA): Soil Science for Global Food Security

Dr J. Scott Angle, a contemporary agricultural leader, has advanced the role of soil science and sustainable nutrient management in ensuring food security. Through leadership roles in the U.S. Department of Agriculture, International Fertiliser Development Centre, and universities, he has promoted sustainable intensification—growing more food with fewer resources. His philosophy of evidence-based policies emphasises the importance of translating scientific knowledge into action to feed a growing world population.

11.2 From Tools to Consciousness

The arc begins in 18th-century England, where Jethro Tull gazed at the chaos of fields scattered with hand-sewn seed and imagined order. His invention of the seed drill in 1701 mechanised planting, ensuring precision and uniformity. Farmers who once tossed seed by hand now watched straight rows of wheat emerge like soldiers in formation. Many ridiculed him at first—"fiddling with gadgets," they said—but when yields improved, scepticism gave way to adoption. Tull's drill was more than a machine; it was the dawning consciousness that farming could be engineered, planned, and rationalised.

Seed Drill

The first machine to sow seeds at the correct depth and spacing—transforming farming from guesswork to an engineering discipline.

Around the same time, in Sweden, Eva Ekeblad turned to a crop still viewed as exotic and even suspicious: the potato. Grain shortages were pushing peasants to hunger, and Ekeblad's research revealed that potatoes could be turned into flour and alcohol, freeing wheat and barley for bread. Soon, bread was on tables that would otherwise have gone empty. When the Swedish Academy inducted her as its first female member, they praised her not as a countess, but as a scientist who had fed her people. Her contribution showed that agricultural breakthroughs were not only technical but profoundly humanitarian.

Potato Flour

In the 1740s, her innovation transformed the potato from curiosity into a staple food and famine shield.

11.3 The Genetics of Destiny

In the quiet garden of a Moravian monastery, Gregor Mendel bent over rows of peas, making meticulous crosses. The neighbours called him "the monk in the garden," unaware that his notes were mapping out the laws of heredity. He counted pods late into the night, patiently tallying wrinkled versus smooth, green versus yellow. Though ignored in his lifetime, his insight—that traits followed predictable laws—was a revelation waiting to change agriculture forever.

Pea Experiments

29,000 pea plants, countless cross-pollinations— yielding the laws of inheritance.

Decades later, Erich Tschermak, rediscovering Mendel's forgotten work, walked through his own experimental plots in Austria. There, he noticed something extraordinary: when two varieties were crossed, the offspring often outperformed both parents. This "hybrid vigour" would become a cornerstone of modern breeding. Farmers who adopted hybrid maize or wheat soon saw harvests heavier than ever before. Tschermak not only confirmed Mendel's work but also translated his science into practical applications in the granary.

Hybrid Vigour

The phenomenon where cross-bred plants yield more than either parent, forming the basis of modern hybrid seeds.

11.4 Soil, Spirit, and Social Renewal

In India, Sir Albert Howard rode his horse through villages and observed how farmers treated soil as sacred. He studied their composting methods and perfected the Indore Process, a systematic approach to making humus. In his diary, he wrote with awe: "The peasants of India are my professors." His book, An Agricultural Testament, warned the industrialising world that to rob the soil was to rob the future. Today's organic and regenerative farming movements trace their lineage back to his fields in India.

The Indore Process

A composting method developed in the 1920s in Madhya Pradesh, the root of the organic farming movement.

Across the ocean, George Washington Carver became a symbol of how agriculture could restore not just land, but people. At Tuskegee Institute, he would gather farmers in simple classrooms and show them jars of peanut-based products—flour, oil, even ink. "God gave us these," he would say, urging them to diversify beyond cotton and replenish exhausted soils. Born enslaved, Carver never sought riches; he often gave away his formulas freely. Once, when Henry Ford offered him wealth for his innovations, Carver declined, insisting his mission was to uplift the poor farmer. His life revolved around agriculture, which he viewed as a means of emancipation.

Agro Products

Craver developed over 300 peanut-based products—from cosmetics to paints—yet gave away many ideas freely.

11.5 The Green Revolution and Beyond

The 20th century brought crises of hunger on a global scale. In the wheat fields of Mexico, Norman Borlaug often worked until dusk, his trousers soaked from irrigation water, his notebook filled with observations of rust-resistant strains. Local farmers mocked him—"too many varieties, too much meddling"—but when his dwarf wheats yielded two to three times more grain, governments took notice. His varieties spread across India and Pakistan, averting famines predicted to claim millions of lives. In 1970, when awarded the Nobel Peace Prize, Borlaug humbly said: "You can't build peace on empty stomachs."

HYV

Dwarf wheat varieties introduced in India doubled yields in the 1960s and 1970s, saving millions from famine.

Meanwhile, in China, Yuan Longping dreamed of a rice field tall enough to shelter him from the sun. In his laboratory, he laboured to create the world's first high-yielding hybrid rice. His breakthrough in the 1970s doubled rice yields, feeding hundreds of millions. Visitors to his research farm would often find him wading barefoot in the paddies, chatting with farmers. Asked once why he never retired, Yuan replied: "As long as there are people who still go hungry, I will keep working."

Hybrid Rice

By the 1990s, it was feeding 60% of China's population—perhaps the single greatest hunger-reduction achievement in history.

11.6 Contemporary Stewards

Today's leaders carry forward the same blend of science and humanity. William Dar, as Director General of ICRISAT, often walked through the parched villages of the Sahel and South Asia, listening to smallholders. His model of "inclusive market-oriented development" argued that technologies must not bypass the poor. As Agriculture Secretary of the Philippines, he pushed for climate-resilient crops and farmer cooperatives, declaring: "Agriculture must be science-led and farmer-centred."

IMOD

Inclusive Market-Oriented Development—a framework to ensure smallholders are not left behind by innovation.

In the United States, Dr J. Scott Angle was a champion of soil science and sustainable nutrient management. In one lecture, he held up a handful of soil and told students: "This is the most important technology we have." His work in universities, the USDA, and the International Fertiliser Development Centre underscores a central truth: without healthy soil, there is no future food security.

Soil as Technology

A living system of microbes, minerals, and organic matter—humanity's oldest and most irreplaceable "infrastructure."

11.7 A Living Legacy

From Jethro Tull's seed drill to Yuan Longping's rice paddies, from Carver's peanut jars to Borlaug's rust-resistant wheats, these leaders tell a single story: agriculture is civilisation's most enduring enterprise. They differ in tools and methods—mechanisation, genetics, soil care, hybrids, policy—but they converge on a truth: farming is not only about calories, but about consciousness.

As we now stand in a tech-driven world—where artificial intelligence, biotechnology, and climate adaptation promise to reshape farming—their legacy calls out to us. Technology can amplify yields, but only leadership infused with ethics, compassion, and inclusivity can secure the future of food.

The torch of agricultural consciousness has passed through their hands to ours. In the fields of tomorrow, farmers will not only harvest grain; they will cultivate dignity, resilience, and hope. The story of agriculture is the story of humanity itself—and its future depends on leaders who dare to see farming as both science and soul.

11.8 Consciousness as the Path of Evolution

Human progress has never advanced solely through self-interest. The greatest leaps in civilisation have come when individuals lifted their work from the plane of personal gain to the realm of universal welfare. To act not merely for oneself but for the countless unseen, the yet unborn, is the essence of higher consciousness. This is how evolution unfolds—not as the survival of the strongest individual, but as the survival of the most compassionate vision.

The ten leaders we have traced in this chapter remind us of this truth. Jethro Tull did not build his seed drill to enrich himself alone; he sought to multiply the bread of his nation. Eva Ekeblad's potato flour was not an invention for the elite, but for peasants threatened by hunger. Mendel, in his quiet monastery, never knew his experiments would one day feed billions. Carver gave away his formulas freely, Borlaug toiled in dusty Mexican fields for farmers he would never meet, and Yuan Longping spent his life in rice paddies dreaming of a world without famine. Dar and Angle, in their policies and soil stewardship, continue to labour for farmers they will never personally know.

Each of them lived a paradox: their impact touched the lives of billions, yet most of those individuals would never know their names. This anonymity is the very sign of greatness. They acted not for applause but for posterity. Their legacy proves that when human beings raise their work to the level of consciousness where the larger welfare of humanity matters more than personal gains and losses, then civilisation takes a step forward.

If mankind fails to make this leap—if it remains trapped in narrow self-interest—it will wither, for no civilisation can endure when its food, soil, and sustenance are managed without care for the whole. But when even a few visionaries embody higher consciousness, they light the path for all.

The future of food, and indeed the future of humanity, rests upon this principle: our survival depends on the capacity to serve beyond the self. These ten leaders did not merely cultivate crops; they cultivated consciousness itself. In their work lies the pattern of evolution, and in their legacy lies humanity's hope of enduring.

11.9 The Higher Call of Agriculture

Civilisations do not collapse for want of riches; they collapse when wisdom is abandoned. True progress cannot be measured by the wealth of a few, but by the well-being of the many—by how a community learns to endure hardships, to flourish in harmony with nature, and to pass forward both sustenance and spirit to generations yet unborn. The ten great leaders of agriculture remind us that greatness lies not in amassing fortune or acclaim, but in lifting one's work to a higher plane of consciousness—where the plough is guided by responsibility, and the harvest is meant for the nourishment of all humanity. They sowed more than seeds of grain; they sowed seeds of justice, dignity, and compassion. Their true harvest was not weighed in markets, but in lives saved from famine, in hope rekindled where hunger once prevailed, and in the promise of food security for billions yet unseen. The ultimate measure of a civilisation is not what it builds for itself, but what it preserves and passes on to those it will never see.

A mind all logic is like a knife all blade.
It makes the hand that uses it bleed.
Gurudev Rabindranath Tagore

History is unflinching in its verdict: work born of self-interest withers, but work done for the larger good endures. A civilisation that clings only to profit and possession signs its own death warrant; but one that dares to serve others, as these visionaries did, writes itself into the future. Humanity's destiny is not secured by competition alone, but by consciousness—by the ability to recognise that survival is not of the strongest, but of the most mindful stewards of life. The soil beneath our feet and the seeds in our hands are not mere instruments of production; they are sacred trusts, carrying within them the story of our survival and the hope of our evolution.

Let this be our inheritance: to till not only the land, but also the human spirit; to cultivate not only crops that feed the body, but also compassion that nourishes the soul. Let us aim to leave behind not merely overflowing granaries, but overflowing reserves of justice, care, and wisdom—treasures that no drought or famine can exhaust. For true wealth is not measured in yields alone, but in the dignity of labour, the fairness of distribution, and the harmony we preserve with the earth that sustains us.

In doing so, we walk in the footsteps of those visionaries who lifted agriculture beyond toil and transaction—transforming it into civilisation's bedrock, where survival gave way to service, and food became more than sustenance: it became solidarity. They understood that a ploughed field is not just a means of livelihood, but a covenant with life itself.

And so, the enduring truth remains: to work for life beyond the self is not only the noblest path, it is the only path by which life itself endures. It is here that agriculture ceases to be an occupation and becomes a sacred duty, a testament that human progress is secure only when it safeguards the unseen billions yet to come. As Mahatma Gandhi reminded us: "The world has enough for everyone's need, but not enough for everyone's greed."

Summary

This chapter traces the lives and legacies of ten global agricultural pioneers—Jethro Tull, Eva Ekeblad, Gregor Mendel, Erich Tschermak, Sir Albert Howard, George Washington Carver, Norman Borlaug, Yuan Longping, William Dar, and J. Scott Angle—who transformed farming from a mere occupation into a civilisational force. Through innovations ranging from the seed drill and hybrid crops to organic farming and soil stewardship, they demonstrated that agriculture is not just about producing food, but also about sustaining societies, empowering the poor, and serving future generations. Their work, often anonymous and selfless, improved the lives of billions they would never know. The chapter argues that true evolution occurs when human endeavour rises beyond individual gain toward the larger welfare of humanity, without which mankind risks perishing. It concludes that agriculture, when guided by higher consciousness, becomes both the foundation of survival and the pathway of hope for the future.

Academic Exercise

1. What farming tool did Jethro Tull invent, and why was it significant?
2. How did Eva Ekeblad's work with potatoes help prevent famine?
3. What key laws did Gregor Mendel discover through pea plant experiments?
4. What was Sir Albert Howard's main message about soil and farming?
5. Which two scientists are remembered as "fathers" of the Green Revolution and hybrid rice?
6. What common quality united all ten agricultural leaders described in the chapter?

12

Self-Reliance

> **The track is long, and the finish line is unseen. But so long as each runner gives their best, humanity moves forward. The baton now rests in your hands. Run your race, and leave the track stronger for those who follow.**

12.1 India's Agricultural Renaissance

Agriculture in any country does not flourish by chance. It requires the deliberate convergence of forces—visionary political leadership that recognises farmers as the backbone of the nation, laws and policies that safeguard land and livelihoods, scientists and reformers who transform knowledge into practice, and above all, a robust education and extension system that ensures even the smallest farmer is not left behind. Where any one of these is missing, agriculture falters; where all align, it becomes the foundation of national strength.

India's own journey from scarcity to abundance is among the greatest success stories of the 20th century. At independence, the nation was scarred by memories of famine, poverty, and dependence on imported food grains. Many predicted that India, with its vast and growing population, would forever remain vulnerable to hunger. Yet within two generations, the country reversed this destiny—becoming a food-surplus nation, a major exporter of staples, and a leader in cooperative dairying and agricultural research. This was no accident. It was the outcome of courage, sacrifice, and foresight by extraordinary individuals who refused to accept dependency as India's fate.

12.2 The Foundations: Self-Reliance and Conviction

The leaders who shaped India's agricultural renaissance were united by three beliefs. First, India must be self-reliant in food. Dependence on others for survival was not merely an economic risk, but a compromise of sovereignty. Second, that India's traditional knowledge carried immense wisdom—from seed preservation to soil care—which must be preserved and adapted, not discarded. Third, true progress requires fearlessness—the courage to experiment with new technologies, learn from failures, and create institutions capable of outlasting individual lives. It was these convictions that turned India's vulnerability into resilience.

Unlike industries where one model can be scaled across geographies, agriculture is inherently local. Soils differ not just in fertility but in texture and chemistry. Water is unevenly distributed. Pest pressures shift with climate. Cropping patterns are shaped by cultural preferences, festivals, and diets that are unique to each region.

Thus, there are no universal global solutions to agriculture. A wheat hybrid that thrives in Punjab may fail in Bihar's soils. A cooperative dairy model that succeeded in Gujarat required adaptation to work in Karnataka. India's genius lay not in copying

Western models but in inventing and indigenising—taking global science, blending it with local conditions, and creating homegrown solutions.

The Land Reforms Era: Leaders like **Govind Ballabh Pant** and **Charan Singh** recognised that political freedom was incomplete without agrarian reform. Their push for tenancy rights and equitable land distribution gave farmers a sense of dignity and ownership.

Institution-Building: Visionaries like **Panjabrao Deshmukh** laid the groundwork for agricultural universities and cooperative movements, embedding science in rural life.

The Green Revolution: When famine loomed in the 1960s, leaders like **C. Subramaniam**, scientists like **M. S. Swaminathan**, and policymakers like **Balram Jakhar** united to introduce high-yielding varieties, fertilisers, and irrigation. Guided by scientists like **B. P. Pal** and **N. S. Randhawa**, India doubled its cereal output and silenced the spectre of mass hunger.

The White Revolution: **Varghese Kurien** mobilised millions of small dairy farmers into cooperatives, transforming India into the world's largest milk producer. It was not just a revolution of milk but of rural empowerment, especially of women.

Diversification and Horticulture: Scientists like **K. L. Chaddha** and **H. P. Singh** shifted attention beyond cereals to fruits, vegetables, and floriculture, enriching both diets and farm incomes.

The Biotechnology Era: Leaders like **E. A. Siddiq** and **R. S. Paroda** introduced molecular tools and hybrids, while **Surinder Vasal**'s Quality Protein Maize improved nutrition. **R. B. Singh** and **S. Nagarajan** worked tirelessly to conserve natural resources and protect crops from disease.

Grassroots Movements: Thinkers like **G. Nammalvar** reminded the country of organic wisdom, showing that sustainability and profitability need not be opposites.

The New Century: Economists like **Ramesh Chand** brought policy focus to farmer incomes and sustainability, while scientists like **Himanshu Pathak** represent a new generation ready to harness biotechnology and AI for climate resilience.

12.3 India's Lesson to the World

What India achieved was not a single revolution, but a continuum of innovations—land reforms, Green and White revolutions, institution-building, diversification, and grassroots renewal. At each step, progress was achieved by aligning political will, scientific innovation, and the active participation of farmers themselves. India's story carries a lesson for the world: agriculture must always be rooted in local realities. Global knowledge is valuable, but it must be adapted, tested, and blended with local wisdom. Policies cannot be imported wholesale; they must emerge from the soil and culture of people.

We now present the lives and contributions of 20 individuals who, in their own way, carried the baton of India's agricultural journey. Some were politicians who wrote laws, some were scientists who bred seeds, some were administrators who built institutions, and some were grassroots leaders who walked from village to village.

12.4 The Legacy

Indian agriculture carries a legacy shaped jointly by visionary political leaders who safeguarded the nation's food dignity and scientific stalwarts who transformed fields into living laboratories, ensuring that India could feed itself and inspire the world.

Govind Ballabh Pant (1887–1961)

A freedom fighter turned statesman, Pant believed that political freedom was incomplete without agrarian reform. As Chief Minister of Uttar Pradesh, he pushed land reforms that curbed feudal practices, giving farmers dignity and ownership. His vision laid the groundwork for equitable rural development.

Dr Panjabrao Deshmukh (1898–1965)

India's first Union Agriculture Minister, Deshmuk, came from a farming background. He established rural institutes and cooperatives, convinced that self-help and education were the best pathways for farmers. His efforts institutionalised agricultural education, seeding a scientific temperament in villages.

Choudhary Charan Singh (1902–1987)

Known as the "Champion of Indian Peasantry," Charan Singh rose from humble origins to become Prime Minister. His writings and policies stressed that India's economy was agrarian at its core. He fought tirelessly for land reforms, farmer-friendly credit, and fair market structures.

C. Subramaniam (1910–2000)

As Agriculture Minister during India's darkest food crisis in the 1960s, he showed rare boldness. He introduced high-yielding wheat and rice varieties, along with fertilisers and irrigation support. With political will and scientific partnership, he became the architect of the Green Revolution.

Dr Balram Jakhar (1923–2016)

A visionary parliamentarian and Union Agriculture Minister, Jakhar modernised India's extension services and rural credit systems. He recognised that technology meant little without being carried to farmers' doorsteps, and he built the bridges between laboratories and fields.

Dr Ramesh Chand (b. 1956)

An agricultural economist of today, Chand shapes policies for a future of sustainable and remunerative farming. He emphasises crop diversification, environmental safeguards, and income security, carrying forward the reformist legacy of his predecessors in a new global era.

Dr Benjamin Peary Pal (1906–1989)

Born in Punjab, Pal became the first Indian Director of ICAR. A master plant breeder, he developed wheat varieties adapted to Indian conditions and institutionalised agricultural research. He believed science must be indigenised to be effective.

Dr A. L. Fletcher (1909–1974)

An English scientist who served India with dedication, Fletcher guided plant breeding programs at IARI. His work strengthened cereal breeding and laid the foundation for Indian scientists to carry the torch.

Dr Varghese Kurien (1921–2012)

The "Milkman of India," Kurien, transformed India from a milk-deficient country into the world's largest producer through Operation Flood. By linking science with cooperative structures, he empowered millions of dairy farmers, especially women.

Dr M. S. Swaminathan (1925–2023)

Hailed as the "Father of the Green Revolution," Swaminathan combined genetics, ecology, and policy. His introduction of semi-dwarf wheat and rice changed India's destiny. Yet, he also warned of sustainability, advocating an "Evergreen Revolution" that balances productivity with ecology.

Dr N. S. Randhawa (1927–1996)

As an educator and administrator, Randhawa built India's agricultural university system. He believed trained minds were as important as seeds and irrigation, creating institutions that continue to serve farmers.

Dr K. L. Chaddha (1936–2025)

Known as the "Father of Horticultural Revolution in India," Chaddha expanded India's agricultural horizon beyond cereals. He championed fruits, vegetables, and floriculture, connecting farming to nutrition and global markets.

Dr E. A. Siddiq (b. 1937)

A rice geneticist, Siddiq worked on Basmati and hybrid rice, improving India's global competitiveness in rice exports. His work in molecular genetics brought precision to crop breeding.

Dr G. Nammalvar (1938–2013)

A spiritual farmer and teacher, Nammalvar dedicated his life to organic and natural farming. He walked from village to village, inspiring farmers to trust nature's wisdom and reduce dependence on chemicals. His work blended tradition with ecological science.

Dr Surinder Vasal (b. 1938)

At CIMMYT in Mexico, Vasal co-developed Quality Protein Maize (QPM). Though his career was global, his innovation reached India, enriching the nutrition of millions. He proved that agricultural science is also a fight against hidden hunger.

Dr R. B. Singh (b. 1940)

A plant geneticist and policy leader, Singh advocated conservation agriculture and resource-efficient farming. His voice was crucial in ensuring India's growth did not come at the cost of soil and water.

Dr R. S. Paroda (b. 1942)

As Director General of ICAR and Chair of the Global Forum on Agricultural Research, Paroda brought biotechnology and global partnerships into Indian agriculture. He positioned Indian science on the world stage.

Dr Subrahmaniam Nagarajan (1945–2016)

A wheat pathologist, Nagarajan, shielded India from rust epidemics through resistance breeding. His vigilance protected food security against invisible threats.

Dr H. P. Singh (b. 1950)

An architect of India's horticultural expansion, Singh led the National Horticulture Mission. He diversified India's farming basket to bananas, mangoes, and floriculture, boosting farmers' income and exports.

Dr Himanshu Pathak (b. 1965)

Among the younger generation, Pathak pioneered the use of plant biotechnology and molecular tools for enhancing stress tolerance. Her work exemplifies the next frontier: preparing Indian agriculture for climate change.

Each ran their part of the race, then passed the baton forward. It is like a baton to the next generation—you, the young students, scientists, entrepreneurs, and policymakers who will decide the future of food, farming, and rural life. Your task is not to imitate but to innovate, not to parasitise but to regenerate, not to look only at profit but to safeguard people and planet together. The relay continues. The track is long, and the finish line is unseen. But with every honest step, humanity moves forward.

12.5 Threads That Bind

Looking across these 20 lives, a pattern emerges. Leaders like Pant, Deshmukh, and Charan Singh gave farmers dignity through laws. Policy architects like Subramaniam and Jakhar created enabling frameworks. Scientists from Pal to Swaminathan and Paroda armed India with seeds and technologies. Visionaries like Kurien and Nammalvar showed that science must serve people and ecology alike. New-age thinkers like Ramesh Chand and Himanshi Pathak are ensuring sustainability in the 21st century.

Lessons from India's Journey

Self-Reliance First - From Pant to Swaminathan, leaders understood that India must feed itself.

Science with Soul - Kurien, Vasal, and Nammalvar demonstrated that innovation is not only about yield, but also about people, nutrition, and ecosystems.

Institutions Matter - ICAR, agricultural universities, cooperatives, and extension services proved that knowledge must be carried to the last mile.

Local Solutions, Global Vision - Every soil, crop, and community is unique. Policies and technologies succeeded because they were localised, not imposed.

The twenty figures celebrated here represent more than individual brilliance; together, they form the DNA of India's agricultural transformation. Their legacies remind us that agriculture is not just about food, but about civilisation itself—linking science, politics, ecology, and human dignity.

As the world faces climate change, biodiversity loss, and nutritional insecurity, India's experience holds a lesson: progress comes when leadership is honest, laws are enabling, science is fearless, and education empowers every farmer. Agriculture will always remain a local phenomenon—but with leaders like these, local wisdom can illuminate the global stage.

India's agricultural triumph rests not on importing universal solutions but on tailoring knowledge to local soils, climates, and communities. What worked in Punjab's wheat belt was not identical to Tamil Nadu's rice fields or Maharashtra's dairying villages. Yet, through courage and leadership, India demonstrated to the world that self-reliance is possible—even in the face of immense challenges.

The lives of these 20 people remind us that agriculture is not just about feeding stomachs—it is about nurturing dignity, sustaining cultures, and ensuring civilisations thrive. Their legacy is a seed for the future: that honest leadership, fearless science, and deep respect for local wisdom can secure food for all.

12.6 Guardians of the Future

Dear Young Scientists, Innovators, and Entrepreneurs, you stand today at the threshold of a transformation greater than any your predecessors have known. Just as the Green Revolution turned the tide of famine into food security, the tide now rising is Artificial Intelligence. It promises to amplify knowledge, sharpen precision, and unlock possibilities that previous generations could not even imagine. Yet, like all great forces of change, it comes with a warning: the same tide that can lift a nation can also sweep away its values, its biodiversity, and the very balance on which life depends.

Agriculture is unlike any other sector—it touches not just markets, but also livelihoods; not just yields, but human dignity. It is woven into the culture of our villages and the health of our ecosystems. To treat it as merely a business of profit, or worse, to let it become parasitic on the environment, is to betray the trust of millions who depend on it. Your mission, therefore, is not only to innovate but to safeguard—to ensure that every breakthrough strengthens the farmer, enriches the soil, and respects the biodiversity that carries within it the seeds of future medicines, cures, and resilience.

AI is not your master; it is your tool. Let it predict weather patterns, optimise irrigation, detect pests, and streamline supply chains. Let it connect a farmer in a remote hamlet to global markets. However, do not let it replace the farmer's wisdom, nor make agriculture a hostage to algorithms that prioritise efficiency over humanity. Ride the tide with intelligence and restraint—steering it toward equity, resilience, and sustainability. Remember, technology is blind unless guided by conscience.

In the rush of urbanisation and infrastructure, it is easy to bulldoze forests and drain wetlands. But these are not wastelands—they are vaults of living knowledge. Within them lie the genes that will protect crops from future pests, the compounds that will heal diseases yet unknown, and the balance that sustains our climate. If biodiversity is lost, the future of agriculture itself is lost. As custodians of the land, you must fight to preserve this wealth, for it is a gift not to us alone but to generations yet unborn.

Pause for a moment, and imagine yourself not just as a scientist, entrepreneur, or business leader—but as an agent of God's will on Earth. You have been entrusted with the power to shape how humanity eats, survives, and thrives. You are asked not merely to innovate but to elevate—to make this planet more habitable, more beautiful, more just. Every choice you make—every crop you breed, every business you build, every technology you deploy—echoes into the future.

Picture an India where farmers utilise AI-powered tools while retaining sovereignty over their seeds; where agribusiness thrives without compromising soil fertility; where cities expand without harming forests; where profits are generated, but not at the expense of justice. This is the tomorrow you are called to create—not as dreamers alone, but as doers who carry both courage and conscience.

The tide of AI will not wait. It will rise whether we fear it or embrace it. Your task is not to resist it, nor to be swept away by it, but to navigate it wisely—anchored in ethics, driven by innovation, and devoted to the welfare of both people and planet. Walk forward as visionaries who remember the farmer, as scientists who honour the soil, and as entrepreneurs who know that true wealth is measured not in profits but in the trust of the people and the health of the Earth.

12.7 Manifesto of the 21st Century Agriculture Leader

The Manifesto of the 21st-Century Agriculture Leader is a bold blend of blueprint and roadmap, charting a future where science, technology, ethics, and farmer wisdom converge to build resilient, regenerative, and inclusive food systems for a changing world.

1. Remember Agriculture is Sacred

Agriculture is not merely an economic activity but the foundation of human civilisation. It feeds, employs, and sustains life. Treating agriculture as sacred means respecting the farmer, the soil, the seed, and the water as gifts entrusted to us, not commodities to be exploited. Leaders must approach every decision with reverence for this sacred cycle, ensuring that growth does not come at the cost of its sanctity.

2. Be Masters of AI, Not Slaves to It

Artificial Intelligence offers tremendous potential to transform agriculture—predicting weather, improving soil health management, connecting farmers to markets, and increasing efficiency. Yet, leaders must ensure that AI remains a tool guided by human values, not an overlord dictating choices. Mastery means using AI to empower smallholders, not displace them; to protect the environment, not accelerate its exploitation.

3. Protect the Poor and Vulnerable

Agriculture is the livelihood of millions of small and marginal farmers who remain vulnerable to debt, climate shocks, and market volatility. A true leader ensures that technological innovations and policies safeguard their rights, income, and dignity. Protecting the vulnerable is not charity but justice, for they are the hands that feed the nation.

4. Let Biodiversity Be Your Bank

The genetic wealth of seeds, plants, soils, and forests is a priceless bank of resilience. Every crop variety, every medicinal herb, every microorganism carries within it the potential for cures, climate adaptation, and food security. A leader must guard biodiversity with vigilance, resisting monocultures and urban encroachment, for once lost, this bank cannot be rebuilt.

5. Build Without Becoming Parasitic

Agriculture and agribusiness must create wealth, but wealth that endures rather than exploits—never draining the environment or society for short-term gain. A parasitic system that depletes soil, exhausts water, and impoverishes farmers ultimately destroys its own foundation. True leadership must instead pursue regenerative growth, where farming replenishes what it takes, businesses share prosperity fairly, and each act of cultivation strengthens the ecosystem. In such a vision, profit is not extracted at the cost of life but emerges as a dividend of care, responsibility, and renewal—leaving the land more fertile, the water cleaner, and the people more empowered for generations to come.

6. Think Local, Act Global

Soils, climates, and cultures vary across regions; therefore, solutions must always begin locally. Yet, in the interconnected world of the 21st century, local innovations must also find global relevance. Leaders must strike this balance—preserving the uniqueness of local practices while leveraging global markets and technologies to enhance their impact.

7. Innovation with Conscience

Science without conscience can become dangerous. Every new seed, chemical, or digital tool must be tested not only for efficiency but also for its ethical and ecological impact. Innovation must serve humanity, not exploit it. A conscientious leader measures progress not just by yield alone, but also by fairness, inclusivity, and sustainability.

8. Balance Progress with Preservation

The march of progress must never bulldoze nature. Roads, dams, cities, and industries are necessary, but they must be planned in harmony with forests, rivers, and wildlife. Preservation is not an obstacle to development, but its insurance—ensuring that growth today does not compromise survival tomorrow.

9. See Yourself as an Agent of God

A leader in agriculture carries an almost divine responsibility: to feed humanity and sustain life. To see oneself as an agent of God is to act with humility, courage, and stewardship. Every action—

whether policy, research, or entrepreneurship—must be guided by the awareness that one is serving a purpose larger than personal gain.

10. Sow Seeds of Trust

At the heart of leadership lies trust—between scientists and farmers, between policymakers and citizens, between humans and nature. Just as seeds sprout only in fertile soil, progress thrives only where trust exists. Leaders must cultivate honesty, transparency, and fairness so that society moves forward together, confident that no one is left behind.

12.8 Conclusion: Carrying the Baton

Life, in its deepest rhythm, is like a great relay race. Each generation runs its segment of the track, carrying in hand not just the weight of years lived but the distilled wisdom of those who came before. A young student opens a book, and in that act, touches the thoughts of an earlier runner—leaders, scientists, dreamers, farmers—who left behind not merely words but experiences, insights, and struggles transformed into lessons.

As that student grows, learning takes the shape of action. The ideas once read become choices in laboratories, villages, boardrooms, and fields. Mistakes are made, resilience is tested, and new knowledge is created. What was once inherited becomes thought, and then it is embodied as work. And when the time comes, this too is written down, shared, and passed on to the next runner.

This book is such a baton. It carries within its pages the voices of those who ran before you—their victories and their defeats, their courage and their doubts, their experiments and their sacrifices. It contains not only their scientific breakthroughs and policy reforms but also the deeper spirit that guided them: a belief that agriculture is not simply about producing food but about nurturing life, sustaining civilisations, and preserving the fragile harmony between humanity and nature.

They ran their race in times of famine, in eras of scarcity, in moments of doubt when the very survival of millions was at stake. Some built institutions, some sowed seeds, some shaped laws, and others walked from village to village, carrying knowledge in their hands and hope in their hearts. Each of them knew that their work would not be completed in their lifetime—that the future belonged to those who would come after.

Now, the baton is in your hands. To you, the young generation, falls the responsibility of carrying forward this legacy in an age of unprecedented possibilities and profound challenges. Artificial Intelligence, biotechnology, climate science, and digital networks will all be at your command. Yet tools alone are not enough; it is integrity, imagination, and compassion that will define how far and how well you run your part of the race.

Run not for yourself alone, but for the farmers whose hands still till the soil with patience and hope, often unseen yet indispensable. Run for the ecosystems that cradle hidden futures—the forests that guard our waters, the soils that harbour unseen life, and the seeds that hold tomorrow's harvest. Run for the billions who depend on agriculture not only for sustenance, but for dignity, security, and the continuity of civilisation itself.

Know that this race is not towards a single finish line, for agriculture has no end—it is a continuum of stewardship, where each generation crosses a milestone and leaves the path stronger for those who follow. The responsibility is vast, but so is the privilege: to be a custodian of life itself.

For agriculture is not merely the art of growing food—it is the art of growing civilisation. To dishonour it is to weaken the very roots of humanity; to nurture it is to ensure that the tree of life stands tall, offering shade, fruit, and hope to all. The baton is now in your hands—run with integrity, courage, and imagination, and let the future remember you not just as a runner, but as a builder of civilisation.

Summary

And when your part of the race is done, leave behind not exhaustion but inspiration, not depletion but renewal. Pass the baton, strengthened by your integrity and enriched by your imagination, so that those who follow can run with hope. For agriculture is not only the heartbeat of civilisation—it is the thread that binds generations together. And you are now its runner. India's agricultural renaissance is a story of self-reliance—transforming famine into food security through the courage of leaders, scientists, and farmers who aligned vision, policy, and innovation. By indigenising global science to local soils and cultures, India proved that sovereignty begins with food. Twenty pioneers—from Pant and Subramaniam to Kurien and Swaminathan—built institutions, revolutions, and ideas that safeguarded both farmers and civilisation. Today, a new generation inherits this baton in the age of AI, tasked not only to innovate but to preserve dignity, biodiversity, and justice. Agriculture remains humanity's sacred foundation, and its future rests on leadership that balances progress with preservation.

Academic Exercise

1. Why was self-reliance in food central to India's sovereignty?
2. What made India's agricultural journey distinct from that of the West?
3. Which leaders laid the foundations of India's Green and White Revolutions?
4. How does agriculture go beyond feeding people to sustaining civilisation?
5. What lesson does India's experience offer to the world?
6. What responsibility does the new generation carry in the age of AI?

Epilogue

From Yield to Fair Deal

For too long, Indian farmers—the very backbone of the nation's food security—have remained at the bottom of the value chain. While consumers pay one price, farmers receive barely a fifth of it; the rest is consumed by layers of trading, intermediaries, and opaque supply chains. This imbalance has created a cycle of poverty, debt, and despair, leaving those who feed the nation unable to adequately feed themselves.

The future of agricultural leadership cannot be confined to producing higher yields alone. Productivity without prosperity is an illusion. True leadership lies in ensuring that farmers receive a fair deal—one that recognises their labour, rewards their stewardship of the land, and restores dignity to their profession. This means more than incremental reforms; it calls for systemic transformation.

Artificial Intelligence can be the catalyst for that transformation. AI-driven platforms can create transparent, real-time marketplaces where farmers directly connect with consumers and fair-minded buyers. Smart systems can guide them to the best inputs, optimise water and soil resources, and provide accurate forecasts that mitigate climate and market risks. By reducing inefficiencies in logistics and eliminating exploitative intermediaries, AI can ensure that the farmer's share of the consumer's rupee is no longer a fraction, but half—or even more.

But technology alone is not enough. It must be guided by leaders with vision, ethics, and empathy—leaders who see agriculture not as an industry to exploit, but as a lifeline to nurture. To lead agriculture in an AI-driven world is to champion fairness as much as efficiency, and equity as much as innovation.

The real measure of progress will not be in how much food we grow, but in how justly we reward those who grow it. When farmers are empowered with knowledge, protected against risks, and connected to transparent supply chains, agriculture will no longer be a story of survival, but of dignity, prosperity, and hope.

The future of Indian agriculture must be written not only in terms of higher yields, but also in terms of fairer deals.

Index

Acknowledgements

Writing "*The Future of Food*" has been both a professional milestone and a deeply personal journey. This book embodies not only my experiences as a scientist and educator but also the collective wisdom, support, and encouragement of many mentors, colleagues, and friends who have shaped my life and thinking. I extend my heartfelt gratitude to all of them.

I would like to express my deep appreciation to my colleagues at NAARM for their unwavering support, intellectual companionship, and encouragement throughout the conceptualisation and writing of this book. I am particularly grateful to **Dr V. Ramasubramaniam**, **Dr Senthil Vinayagam**, **Dr R. Venkattakumar**, **Dr B. Ganesh Kumar**, **Dr A. Dhandapani**, **Dr Ranjit Kumar**, **Dr K. Srinivas**, **Mr B.P.R. Vithal**, **Mrs A. Mercy**, and **Mrs T. Vanisri** for their valuable insights and collaboration. I also wish to acknowledge with appreciation **Dr Balakrishnan**, **Dr Prem Chand**, **Dr Surya**, **Dr Venkatesan**, **Dr Nirmala**, **Dr Renuka**, **Dr Umesh**, **Dr Ramesh**, and **Dr Suresh**, whose cooperation and goodwill enriched my experience at the Academy and contributed meaningfully to this endeavour.

My sincere thanks go to the visionary leadership and colleagues of the Indian Council of Agricultural Research (ICAR), whose dedication to advancing Indian agriculture continues to inspire generations. I express my profound gratitude to **Dr R.S. Paroda**, **Dr Himanshu Pathak**, **Dr M.L. Jat**, and **Dr H.P. Singh**, whose guidance and encouragement have left an indelible impression on my professional journey.

I am equally thankful to **Dr Rajveer Singh**, **Dr Ch. Srinivasa Rao**, **Dr R.C. Agrawal**, **Dr J. Jena**, **Dr S.L. Mehta**, **Dr Vimlesh Maan**, **Dr S.K. Sharma**, **Dr Ranjit Singh**, **Dr Seema Jaggi**, **Dr Anuradha Agarwal**, **Dr Vinay Bhardwaj**, **Dr P.K. Rai**, **Dr Khem Chand**, **Dr V.K. Singh**, **Dr Tara Satyavathi**, **Dr Shaik N. Meera**, **Dr R.K. Mathur**, and **Dr R.M. Sundaram** for their institutional support and professional collaboration. I also gratefully acknowledge **Dr Jankiram**, **Dr Nirmalbabu**, **Dr A.K. Poonia**, and **Dr P.L. Saran** for their valuable guidance and friendship, which have been instrumental in shaping my work and perspective.

I extend my sincere appreciation to **Dr Sanjay Kumar**, Chairman, **Dr B.S. Dwivedi**, and **Dr Major Singh**, Members of ASRB, for their support, encouragement, and commitment to maintaining excellence in agricultural research and education.

With reverence and gratitude, I bow to my mentors who have shaped my scientific thinking, discipline, and values. I pay homage to my late guides, **Dr N.L. Sen**, **Dr S.K. Jain**, **Dr S.C. Bandari**, **Dr K.C. Pundrik**, and **Dr P.R. Kameria**, whose wisdom and teachings remain alive in my work.

I express heartfelt thanks to **Dr G.R. Chaudhary**, **Dr B.L. Baser**, **Dr P.K. Dashora**, **Dr L.K. Dashora**, **Dr K.K. Bohra**, and **Dr A.K. Purohit** for nurturing my academic foundation. I am equally indebted to my mentors, **Dr P.L. Gautam**, **Dr S.R. Maloo**, **A.K. Singh**, **Dr R.L. Shyampura**, and **Er. S.K. Haldar**, whose visionary guidance and trust have been sources of lifelong motivation.

My heartfelt appreciation goes to my friends and associates who have stood by me with unwavering support and goodwill throughout my professional life. I am thankful to **Dr Dinesh Arora, Dr R.K. Sharma, Dr S.K. Sharma, Dr D.S. Bhati, Dr S.S. Meena, and Dr M.D. Meena, Dr M.K. Vishal, Dr P. Bhatnagar, Dr Dayanand, Dr A.S. Godara, Dr M.M. Kumawat, Dr Bhanwar Lal, Dr A.K. Soni**, and **Dr M.S. Fageria** for their friendship and encouragement. I also wish to specially mention **Dr Deepak, Dr J.P. Tetarwal, Dr Jagdish Lal, Dr M.L. Maheria**, and **Dr Ravindra Kolapur**, whose companionship and discussions added warmth and insight to my journey.

Finally, I express my profound gratitude to **Prof. Arun Tiwari**, my co-author and friend, whose intellectual brilliance, narrative depth, and unwavering commitment helped transform *Future of Food* from an idea into a compelling discourse. His ability to weave science, philosophy, and vision together gave this book its enduring spirit.

Above all, I bow to the **Supreme Power** for guiding me through every step of this journey. I owe heartfelt gratitude to my **family**—whose patience, love, and faith have been my greatest strength. To my parents, who instilled in me the values of hard work and humility; to my wife, who has been my constant companion through every challenge; and to my children, who continue to inspire me to dream for a better tomorrow.

This book is dedicated to the farmers of India—the silent heroes who toil to feed the nation—and to all scientists and innovators working tirelessly for a sustainable, equitable, and nourishing future. *May the pursuit of knowledge always serve humanity, and may the future of food be the future of life itself.*

— Dr Gopal Lal

About the Authors

Dr Gopal Lal is a distinguished horticultural scientist and the Director of the ICAR–National Academy of Agricultural Research Management (NAARM), Hyderabad. With nearly three decades of dedicated service in agricultural research, education, extension, and institutional leadership, he has made pioneering contributions to arid horticulture, seed-spice genetics, crop improvement, organic farming, post-harvest management, and value chain development.

He previously served as Director of the ICAR–National Research Centre on Seed Spices, Ajmer; Member (Agriculture) of the Cauvery Water Management Authority, New Delhi; and Professor of Post-Harvest Management at the College of Horticulture and Forestry, Central Agricultural University, Pasighat. He is also a Member of the Rajasthan Kisan Ayog, Chairman and Member of several Task Force Committees of PPVFRA, and a Member of the Seed Spices Task Force Committee and the Spices Board of India.

A visionary leader and accomplished mentor, Dr Lal has guided 13 MSc and PhD scholars, led 77 research projects—including eight externally funded initiatives—and developed 13 high-yielding crop varieties along with several commercialised technologies. His prolific academic output includes 27 authored, edited, and institutional books, as well as over 200 research publications of international repute.

Recipient of multiple national awards, fellowships, and honours, Dr Lal is widely recognised for advancing sustainable horticultural and agricultural transformation in India. His contributions extend to policy formulation in women's empowerment in agriculture and the strategic promotion of drone technologies for modernising India's farm and allied sectors, reaffirming his role as a transformative leader shaping the future of Indian agriculture.

Prof. Arun Tiwari is a former missile scientist who worked on the hardware of Trishul and Akash Missiles and later led the Civilian Spinoff of the Defence Technology Programme, which included developing indigenous coronary stents in the 1990s and piloting the Pan-Africa e-Network project for the Government of India. He taught Medical Humanities at the University of Hyderabad as an adjunct professor for ten years (2008-2018). He co-authored *Wings of Fire*, the celebrated autobiography of Dr APJ Abdul Kalam, and has since published 30 additional books.

Now retired from professional service, Prof Tiwari continues to devote himself to writing, mentoring, and thought leadership at the intersection of science, technology, and spirituality. He writes regularly for *PanAfrican Agriculture.* His blog (aruntiwari.com), published every fortnight since May 2019, is widely read and has earned him renown for his ability to amalgamate and disseminate knowledge across scientific disciplines and the humanities.

Books by Arun Tiwari

Published by Sakal Publications

- The Baton Should Not Fall
- Decoding the Pandemic
- India Wakes : Post Coronavirus New World Order
- Innovate Locally to Win Globally
- Live for a Legacy
- Uddhava Gita
- Discovery of Anugita
- Simple Spirituality : Recalling Kabir
- A Modern Interpretation of Tulsidas Ramcharitramanas
- A Modern Interpretation of Lokmanya Tilak's Gita Rahasya
- We are What We Eat
- Yoga Vasishtha
- Spectrum
- Purpose of Life
- Neo-ethical Living
- Village Republic 2.0

www.ingramcontent.com/pod-product-compliance
Lightning Source LLC
Chambersburg PA
CBHW051402240926
51852CB00005B/93/J
9789349487277